I0468879

Cutting the Army's Umbilical Cord

A Study of Emerging Fuel Technologies and their Impact on National Security

A Monograph

by

MAJ Matthew A. Price

United States Army

MENS EST CLAVIS VICTORIAE

School of Advanced Military Studies

United States Army Command and General Staff College

Fort Leavenworth, Kansas

AY 2011

Approved for Public Release; Distribution is Unlimited

REPORT DOCUMENTATION PAGE

Form Approved
OMB No. 074-0188

Public reporting burden for this collection of information is estimated to average 1 hour per response, including the time for reviewing instructions, searching existing data sources, gathering and maintaining the data needed, and completing and reviewing this collection of information. Send comments regarding this burden estimate or any other aspect of this collection of information, including suggestions for reducing this burden to Washington Headquarters Services, Directorate for Information Operations and Reports, 1215 Jefferson Davis Highway, Suite 1204, Arlington, VA 22202-4302, and to the Office of Management and Budget, Paperwork Reduction Project (0704-0188), Washington, DC 20503

1. AGENCY USE ONLY (Leave blank)	2. REPORT DATE 15 July 2011	3. REPORT TYPE AND DATES COVERED SAMS Monograph, June 2010 – March 2011	
4. TITLE AND SUBTITLE Cutting the Army's Umbilical Cord A Study of Emerging Fuel Technologies and their Impact on National Security			**5. FUNDING NUMBERS**
6. AUTHOR(S) Major Matthew A. Price, United States Army			
7. PERFORMING ORGANIZATION NAME(S) AND ADDRESS(ES) School of Advanced Military Studies (SAMS) 250 Gibbon Avenue Fort Leavenworth, Kansas 66027-2134			**8. PERFORMING ORGANIZATION REPORT NUMBER**
9. SPONSORING / MONITORING AGENCY NAME(S) AND ADDRESS(ES) Command and General Staff College 731 McClellan Avenue Fort Leavenworth, Kansas 66027			**10. SPONSORING / MONITORING AGENCY REPORT NUMBER**

11. SUPPLEMENTARY NOTES

12a. DISTRIBUTION / AVAILABILITY STATEMENT Approved For Public Release; Distribution Unlimited	12b. DISTRIBUTION CODE

13. ABSTRACT *(Maximum 200 Words)*

The U.S. dependence on foreign oil is a national security concern. The Department of Defense is the largest federal government consumer of oil and the Army plays a significant role in reducing consumption. To do this, the Army must reduce fuel consumption at U.S. installations but most importantly, at deployed locations.

Improving the efficiency and decreasing consumption of sustainment platforms, the largest battlefield consumers of fuel, became an Army priority during Operations Iraqi and Enduring Freedom. This new focus on emerging fuel technologies has the potential to decrease the logistical requirements in theater, reduce the budget outlays for fuel, and reduce risk for soldiers. In order to validate these claims, this monograph analyzes three case studies.

The three emerging fuel technologies evaluated are microgrids, solar and wind power generators, and hybrid-electric tactical wheeled vehicles. The method used in the case studies is to replace an inefficient existing technology with the new one and calculate fuel savings, cost savings, risk reduction, and casualty reduction. The data collected from the analysis of these case studies draw some eye opening conclusions. Most significant is the number of tankers removed from the roads in one-year, which approaches 3,000, corresponding to close to 6,000 soldiers no longer needed in theater to deliver fuel. This decrease of soldiers leads to four soldiers who might have avoided death in Iraq in 2007. Because of these findings, this monograph asserts that the Army use an enterprise approach at developing and implementing emerging fuel technologies in order to decrease fuel cost and risk to soldiers.

14. SUBJECT TERMS Army, Logistics, Emerging Fuel Technologies, Green Power	15. NUMBER OF PAGES 51
	16. PRICE CODE

17. SECURITY CLASSIFICATION OF REPORT UNCLASSIFIED	18. SECURITY CLASSIFICATION OF THIS PAGE UNCLASSIFIED	19. SECURITY CLASSIFICATION OF ABSTRACT UNCLASSIFIED	20. LIMITATION OF ABSTRACT

NSN 7540-01-280-5500

Standard Form 298 (Rev. 2-89)
Prescribed by ANSI Std. Z39-18
298-102

SCHOOL OF ADVANCED MILITARY STUDIES

MONOGRAPH APPROVAL

Major Matthew A. Price

Title of Monograph: Cutting the Army's Umbilical Cord

Approved by:

_____ Monograph Director
Bruce E. Stanley

_____ Monograph Reader
G. Scott Gorman, Ph. D.

_____ Director,
Thomas C. Graves, COL, IN   School of Advanced
                                         Military Studies

_____ Director,
Robert F. Baumann, Ph.D.   Graduate Degree
                                         Programs

Disclaimer: Opinions, conclusions, and recommendations expressed or implied within are solely those of the author, and do not represent the views of the U.S. Army School of Advanced Military Studies, the U.S. Army Command and General Staff College, the United States Army, the Department of Defense, or any other U.S. government agency. Cleared for public release: distribution unlimited.

Abstract

CUTTING THE ARMY'S UMBILICAL CORD, by Major Matthew A. Price, U.S. Army, 51 Pages.

The U.S. dependence on foreign oil is a national security concern. The Department of Defense is the largest federal government consumer of oil and the Army plays a significant role in reducing consumption. To do this, the Army must reduce fuel consumption at U.S. installations but most importantly, at deployed locations.

Improving the efficiency and decreasing consumption of sustainment platforms, the largest battlefield consumers of fuel, became an Army priority during Operations Iraqi and Enduring Freedom. This new focus on emerging fuel technologies has the potential to decrease the logistical requirements in theater, reduce the budget outlays for fuel, and reduce risk for soldiers. In order to validate these claims, this monograph analyzes three case studies.

The three emerging fuel technologies evaluated are microgrids, solar and wind power generators, and hybrid-electric tactical wheeled vehicles. The method used in the case studies is to replace an inefficient existing technology with the new one and calculate fuel savings, cost savings, risk reduction, and casualty reduction. The data collected from the analysis of these case studies draw some eye opening conclusions. Most significant is the number of tankers removed from the roads in one-year, which approaches 3,000, corresponding to close to 6,000 soldiers no longer needed in theater to deliver fuel. This decrease of soldiers leads to four soldiers who might have avoided death in Iraq in 2007. Because of these findings, this monograph asserts that the Army use an enterprise approach at developing and implementing emerging fuel technologies in order to decrease fuel cost and risk to soldiers.

Acronyms

AEPI   Army Environmental Policy Institute

AOR   Area of Responsibility

AMMPS   Advanced Medium Mobile Power Source

BCT   Brigade Combat Team

COB   Contingency Operating Base

CRS   Congressional Research Service

DA   Department of the Army

DLA   Defense Logistics Agency

DOD   Department of Defense

DSB   Defense Science Board

FBCF   Fully Burdened Cost of Fuel

FY   Fiscal Year

GAO   Government Accountability Office

HEMTT   Heavy Expanded Mobility Tactical Truck

HEVEA   Hybrid-Electric Vehicle Experimentation and Assessment

MEP   Mobile Electric Power

MTOE   Modification Table of Organization and Equipment

NSS   National Security Strategy

OEF   Operation Enduring Freedom

OIF   Operation Iraqi Freedom

PM   Project Manager

TQG   Tactical Quiet Generator

U.S.   United States

USCENTCOM United States Central Command

Table of Contents

Introduction

The U.S. imports nearly twelve million barrels of oil per day to meet a strategic demand of twenty-one million barrels of oil per day. This equates to one quarter of the world's total daily demand for oil. The U.S. Department of Defense (DOD) consumes 330,000 barrels per day which accounts for two percent of U.S. oil consumption per day. By far, the US DOD uses more oil than any other US government agency, accounting for ninety-three percent of the governments' energy and oil consumption.[1] The concern is not U.S. vs. DOD consumption but how dependent within the government the DOD is on fuel and oil. With DOD oil consumption eclipsing entire foreign nation's consumption and rivaling countries like Cuba, Denmark, Ireland, and Puerto Rico, the dependence presents a national security concern for the United States. In order to decrease the DOD's dependence, Congress directed that the DOD begin to address the excessive consumption in three ways.[2] First, the Secretary of Defense must include fuel logistic vulnerabilities in force planning and the acquisition decisions. Second, the Secretary must include fuel efficiency as a key performance parameter in the requirements development process. Third, the Secretary must implement the Fully Burdened Cost of Fuel (FBCF) into life-cycle analysis conducted on new capabilities. Most importantly, the Secretary had to comply with these Congressional mandates within three years of enactment.[3]

Once the Act became public law, the DOD subsequently published comments within the Quadrennial Defense Review (QDR) establishing a plan to address the Congressional directives. Within the QDR, the DOD recognized that energy efficiency serves as a force multiplier, since it

---

[1] Deloitte, *Energy Security, America's Best Defense*, 10, http://www.deloitte.com/view/en_US/us/Industries/oil- gas/ (accessed May 24, 2011).

[2] *Duncan Hunter National Defense Authorization Act for Fiscal Year 2009*, S.3001, 110th Cong., *Congressional Record Summary*, The Library of Congress (October 14, 2008), http://thomas.loc.gov/ (accessed May 17, 2011). Sec. 332.

[3] Ibid.

increases range and endurance of forces in the field and reduces the number of combat forces diverted to protect energy supply lines.[4]

It follows that the risk of soldiers killed or wounded while delivering or protecting these energy supply lines presents its own national security implications. According to a recent study, for every 72,300 barrels of fuel consumed in Iraq, insurgents kill one soldier charged with either operating, or protecting fuel convoys.[5] This is an area where emerging fuel technologies can have a profound impact on casualty reduction.[6] Since the delivery and protection of these supply lines depend heavily on the Army, lessening the consumption of the Army directly address Congress' concerns.

In order to illustrate the operational impact of reducing fuel consumption, it is necessary to analyze a recent U.S. conflict and the associated fuel consumption. The availability of data and the ability to closely correspond a researched conflict to future conflicts compelled the author to choose Operation Iraqi Freedom in 2007. The context for this monograph gives the perspective on how and why national security hinges on the Army's deepening dependence on oil. Emerging fuel technologies are the tools used to decrease this dependence, which decreases cost and decreases the risk to logistic and force protection soldiers.

This monograph is timely for two reasons. First, this monograph occurs when rising fuel costs and an economic recession severely limit the DOD's ability to budget operations. For every

---

[4] U.S. Department of Defense, *Quadrennial Defense Review, 2010*, 87, http://www.defense.gov/qdr/images/QDR_as_of_12Feb10_1000.pdf (accessed May 17, 2011).

[5] Army Environmental Policy Institute, *Sustain the Mission Project: Casualty Factors for Fuel and Water Resupply Convoys, Final Technical Report, September 2009*, by David S. Eady, et al., 5, http://www.aepi.army.mil/docs/whatsnew/SMP_Casualty_Cost_Factors_Final1-09.pdf (accessed December 20, 2010).

[6] Emerging fuel technologies – these are fuel saving technologies that may be on the civilian market but have not been vetted through the Army procurement process. Some are also military specific platforms that are being tested and developed such as the HEMTT hybrid-electric tactical truck.

$10 increase in the price of a barrel of oil, DOD's operating costs increase by $1.3 billion.[7] Second, this monograph supports the Congressional directive in the Duncan Hunter National Defense Authorization Act from Fiscal Year (FY) 2009. The act directed that the DOD conduct studies of alternative energy sources aimed at reducing fuel consumption. The reasons were to assess the ability of the emerging technology's to reduce the risk of casualties associated with convoys supplying fuel to forward-deployed locations.[8]

This monograph follows a line of inquiry that determines in what ways and to what extent the Army can use emerging fuel technologies to reduce overall fuel consumption. It appears that dependence on fuel and the inherent operational risk it causes is not transparent to Army senior leaders. If these leaders recognize the risk, why has the Army not already used emerging fuel technologies on a large scale? It appears that leaders recognize that decreased dependence increases soldier safety, operational effectiveness, and national security. As such, the objective of this monograph is to evaluate the capabilities of three technologies focused on where each can have the most profound impact. Selected case studies examine solar and wind power generators, hybrid electric vehicles, and microgrids aim at replacing the heaviest battlefield consumers of fuel: generators and tactical wheeled vehicles.

The primary purpose for this monograph is to illustrate how emerging fuel technologies reduce the loss of life attributed to delivering fuel. The monograph illustrates the potential sweeping affects emerging fuel technologies can have on lowering the possibility of the loss of life when the Army implements these technologies across the enterprise. Even with the extensive research conducted within this field, no study discussed the importance of the widespread use of these technologies. Additionally, this monograph adds to the body of knowledge on how the

---

[7] U.S. Government Accountability Office, Report to the Subcommittee on Readiness, Committee on Armed Services, House of Representatives, *Defense Management: DOD Needs to Increase Attention on Fuel Demand Management at Forward-Deployed Locations,* GAO-09-300 (Washington, D.C.: February 2009), 9, http://www.gao.gov (accessed August 24, 2010).

[8] Ibid., 11.

implementation of emerging technologies applied on a small scale can coalesce into national level implications. Furthermore, this monograph has a practical application to augment the existing research, studies, and reports as well as provide a basis for further research using a multiple technology approach.

In order to clarify the Army usage of terms contained in this monograph, it is necessary to discuss the definitions of four terms contained in both the thesis and hypotheses. The first term, emerging fuel technology, describes a technology with the capability of reducing the amount of fuel consumed from an existing technology. Additionally, the use of this technology in the civilian sector may be widespread, but it has not undergone the Army's research and development or the testing and evaluation phases of procurement. The second term, fuel efficiency, describes the efficiency of a piece of equipment by way of percent increases in efficiency, not in terms of increases in units of gallons per hour or miles per gallon. The third term, fuel consumption, describes the total fuel usage for an entire year assuming a set number of miles driven or a set number of hours used for the existing and emerging technology. The fourth term, fuel cost, refers to the cost of a gallon of fuel using the Defense Logistics Agency (DLA) or the Fully Burdened Cost of Fuel (FBCF) analysis value for the cost of fuel. Both published costs of fuel do not fluctuate daily, as do civilian prices.[9]

There are several limitations in this monograph, which creates the need to make generalizations and inferences in order to conduct the study. Most significant of these is the lack of available data used in the calculations in each case study. Due to the lack of fuel monitoring devices on equipment and the perceived culture of disinterest in fuel consumption, there are no

---

[9] Fully burdened costs include the standard fuel price, direct ground fuel infrastructure, indirect base infrastructure, environmental costs, delivery asset operations and support, delivery asset depreciation, and other specific costs. Source: Army Environmental Policy Institute, *Sustain the Mission Project Report: Energy and Water Costing Methodology and Decision Support Tool: July 2008*, by Steve Siegel, et al., vii, http://www.aepi. army.mil/docs /whatsnew/SMP_Casualty_Cost_Factors_Final1-09.pdf (accessed December 20, 2010).

databases in the Army that capture fuel consumption data for the existing technologies. This hampers real-time data calculations but does not present an insurmountable task at using reasonable approximations based on other secondary sources.

In order to calculate data from the analysis of each case study, which easily aggregates into comprehensible result, the study delimits factors that would change the usage characteristics of the equipment. This includes delimiting weather, changes in usage patterns, repair and maintenance downtime, and changes in weight or electrical load. Additionally, this monograph does not include all Army units deployed to Iraq in FY2007. Two of the three case studies analyze only equipment from brigade combat teams deployed to Iraq in FY2007. This allows the study to focus on units who do not have extreme variations in equipment such as combat service support units.

The above limitations and delimitations lead to necessary assumptions that correctly scope this monograph. The lack of available data discussed above creates the need for this monograph to hold constant fuel consumption factors. An additional assumption is that the Army would implement each technology with minimal variations in capabilities. For instance, if the Army procured solar and wind power generators and used them as replacements for generators in brigade combat teams, these new generators would have very similar power producing characteristics.

The organization of this monograph follows a logical flow of introducing the study, reviewing the available literature used throughout the study, laying out the methodology of conducting the study, analyzing three emerging technologies, and summarizing the data using implications and conclusions. The intent of section two is to provide a literature review of existing research as well as identifying sources, which contain information about the emerging technologies. Section three of this monograph provides a step-by-step guide, which illustrates the author's methods of analysis for each of the emerging technologies and data collection. Section four encompasses the analysis of the three case studies, which focuses on calculating the impacts

5

of the technology on fuel consumption, cost, and risk to soldiers. The final two sections, five and six, consolidate the data from the case studies, discuss implications beyond the scope of this monograph and draw conclusions about the implementation of emerging technologies in the Army.

Literature Review

This section presents the background for researching the relationship between emerging fuel technology and energy consumption of the Army. For five decades, scholars and researchers studied and reported on the rising energy consumption of the military. Their studies reviewed the rising monetary cost of military fuel consumption and national security concerns born from fuel dependence. The following review briefly discusses the early political attempts at policy changes and decreased dependence, but primarily focuses on the last half-decade of pertinent policy changes and fuel reduction research. This monograph seeks to build upon the body of research, augmenting existing dependence studies, through the lens of a practitioner of Army logistics.

This section begins with a review of the literature illustrating how U.S. policy over the last forty years changed how the nation sourced energy. This illustrates how fuel use in the Army attributes to national security concerns. The section continues by illustrating how the Congressional and professional research community, during years of policy changes, produced evidence of the critical need to reduce energy consumption in order to protect national security. This review concludes with solutions created and implemented in 2009 and 2010, showing how four decades of political and DOD policy finally translated into energy reduction solutions for the Army. The intent of this literature review is to show how policy recommendations can help the Army reduce real fuel cost by implementing fuel reduction technologies in order to improve national security.

A National Security Issue

Thomas Kraemer captured how difficult reducing U.S. dependence on foreign energy sources is when attempted with just changes in national policy. He asserts that changing the minds of energy users began over 40 years ago.[10] He believed that in order to embrace new energy sources, the U.S. population, and the Department of Defense required a cultural shift.[11] He supported this argument by discussing the energy policies of successive U.S. Presidential Administrations from Richard Nixon to George W. Bush. The cultural shift began with President Nixon, at a time when the U.S. was in a fuel crisis, when he launched Project Independence in response to the Arab oil embargo of 1973.[12] President Nixon intended to decrease national security vulnerabilities by reducing the demand for foreign supplied energy. To do this, he sought to source domestic energy needs from within the U.S.[13] Furthermore, as Kramer listed each president's attempt at implementing new policies aimed at breaking the nation's addiction to oil, he showed that the hardest part is getting started.[14] He discussed how the last nine presidents mentioned the nation's addiction to oil in their National Security Strategy (NSS).[15] President Obama continued the trend by emphasizing the importance of finding alternative sources of

---

[10] U.S. Army War College, "Addicted to Oil: Strategic Implications of American Oil Policy," COL Thomas D. Kraemer, *Carlisle Papers in Security Strategy*, Carlisle, PA: USA, May 2006, 1.

[11] Ibid., 10.

[12] Ibid., 1.

- In November of 1973, President Nixon launches Project Independence with the goal of achieving energy self-sufficiency by 1980. Nixon declares that American science, technology, and industry can free the United States from dependence on foreign oil. The Project's proposals included: completion of the Trans-Alaska Pipeline for distribution of the oil in the Alaska's North Slope fields; lowering of automobile speed limits to 55 mph; utilization of coal instead of oil in power plants; diversion of funds from highway construction to development of mass transit systems; decontrol of new natural gas; expedition of licensing of nuclear power plants; creation of a U.S. Dept of Energy to promote and oversee energy related projects; funding of a $10 billion research and development program designed to achieve energy self-sufficiency by 1980. Source: Charles E. Brown, *World Energy Resources* (Berlin, Germany: Springer, 2002), 227.

[13] Ibid., 1.

[14] Ibid., 11.

[15] Ibid.

energy in order to decrease cost and dependence in his 2010 NSS. The Obama administration took a broad approach at reducing dependence with national level policy. Their policy asserted that, in continuation with past policies, the nation supports polices aimed at pursuing clean energy technology, better vehicle efficiency, and heavy investment in research.[16]

A recent RAND study suggested that the impact of U.S. dependence on fossil fuels adversely affects the security of the nation.[17] Militarily, the authors asserted two things: one is that the U.S. dependence on energy drives DOD policy on the manning structure of the Army. Second, the study asserted that heavy dependence on imported oil requires substantial military force to maintain the security of international oil flows to the global market.[18] This heavy dependence also requires that the U.S. use military forces to ensure access to oil both for the health of the U.S. economy, and national security.[19] Linking dependence to a military requirement, the study found that the origins of United States Central Command (USCENTCOM) came about as a result of an expanded U.S. military commitment to protect access to Middle Eastern oil.[20] Moreover, the study estimated that the DOD could have saved $67-$83 Billion in FY2008, or 12-15 percent of the U.S. Defense budget, if the DOD did not require the military to secure oil from this region.[21] Finding additional ways to save energy and reduce the need for military actions in that part of the world is one of the driving forces in the continuation of this monograph.

---

[16] Barrack Obama, *National Security Strategy 2010* (Washington, DC: The White House, May 2010), 30, http://www.whitehouse.gov/sites/default/files/rss_viewer/national_security_strategy.pdf (accessed September 24, 2010).

[17] Keith Crane et al., *Imported Oil and U.S. National Security* (Santa Monica, California: RAND Corporation, 2009), 1, http://www.rand.org/pubs/technical_reports/ (accessed September 24, 2010).

[18] Ibid., 59.

[19] Ibid., 60.

[20] Ibid.

[21] Ibid., xv.

The authors of the study *Post-Partisan Power* illustrated the challenge of implementing national policy specifically aimed at reducing dependence on foreign oil. This study highlighted the need to bring the political community together in order to create opportunities for energy innovation and decreased oil dependence.[22] The authors of the study asserted that the U.S. goal should be to make cleaner energy sources cheaper so they can steadily replace fossil fuels.[23] Most importantly, the study asserted that past successes in federal investment, funded through policy changes, successively catalyzed technology innovation by the U.S. military.[24] Additionally, the study suggested placing the challenge of developing new, cheap energy sources directly on the DOD, which has had success when funding allowed for significant focus on energy innovation.[25] The authors concluded their study with statements linking the revitalized use of the DOD in developing solutions to energy dependence to the opportunity to save lives by decreasing the operational demand for fuel.[26]

Operational Issues

Using Congressional research and independent studies to link national policy to field solutions is not new. An independent scientific advisory group conducting research on fuel and national policy since 1960 concluded that there are two key reasons why the DOD should reduce fossil-fuel dependence.[27] The first reason is that transportation of fuel imposes a large logistical

---

[22] Steven F. Hayward et al., *Post-Partisan Power* (AEI Online, 2010), 6, http://www.aei.org/paper /100149 (accessed October 25, 2010).

[23] Ibid., 5.

[24] Ibid., 6.

[25] Ibid., 24.

[26] Ibid., 21.

[27] JASON is an independent scientific advisory group that provides consulting services to the U.S. government and other organizations on matters of defense science and technology. Established in 1960, JASON typically performs studies under contract to the Department of Defense (frequently DARPA and the U.S. Navy) the Department of Energy, the U.S. Intelligence Community, and the FBI. Approximately

burden, which increases operational constraints and vulnerabilities.[28] The second reason is that an uncertain future of world and domestic oil supply increases exposure and vulnerabilities.[29] The study also found that fuel availability over the next forty years would not drive decisions by the DOD to reduce fuel consumption.[30] However, the authors of the study did conclude that the increased costs of delivering fuel to remote locations are the greatest motivation for reduction.[31] Their conclusion is one of the central themes in this paper because a decrease in fuel consumption leads to a reduction in the logistic requirements to support remote locations. Missing in the JASON study, however, is the analysis of the massive amount of fuel required to operate generators at deployed locations.

Additional examples of linking field solutions to national policy surfaced in a report from the Logistics Management Institute, now simply known as LMI.[32] The authors conducted studies on available technologies aimed at reducing consumption and decreasing the logistical burden imposed by using fossil fuels.[33] LMI saw these technologies as opportunities to decrease the DOD's dependence on fossil fuels while decreasing gaps in the DOD's energy strategy. The

---

half of the resulting JASON reports are unclassified. From Federation of America Scientist website: http://www.fas.org/irp/agency/dod/jason/

[28] JASON, *Reducing DOD Fossil-Fuel Dependence, September 2006*, JSR-06-135, iv, http://www.fas.org/irp/agency/dod/jason/fossil.pdf (accessed August 24 2010).

[29] Ibid.

[30] Ibid.

- According to sources in the JASON study on page five, based on economically extractable reserves and a 2005 consumption rate, the world has 41 years of proven oil reserves and it would be cost prohibitive to drill and verify beyond 40 years.

[31] Ibid., 31.

[32] The Logistics Management Institute, now known as LMI, is a non-profit government consulting organization, which began in 1961 following a suggestion from Secretary of Defense Robert McNamara to President Kennedy to create a non-profit research institute focused on achieving major breakthroughs in logistics management where the government spent half of the Defense budget. From LMI's History webpage: http://www.lmi.org/About-LMI/History.aspx

[33] Thomas D. Crowley et al., *Transforming the Way DOD Looks at Energy: An Approach to Establishing an Energy Strategy, FT602T1* (LMI Government Consulting April 2007), iii, http://www.dtic.mil/cgi -bin/GetTRDoc?AD=ADA467003&Location=U2&doc=GetTRDoc.pdf (accessed November 23, 2010).

report cited three gaps, but only two, the operational and fiscal gaps, directly pertain to this monograph. The former defined how the DOD's operational concept seeks greater mobility, persistence, and agility for their forces, but the logistic requirements caused by fuel dependence of these forces limit the DOD's ability to realize the concepts.[34] The latter illustrated why increases in energy demand and increases in the price of energy counter the DOD's pursuit of procuring new capabilities for the future.[35]

A study by the Defense Science Board Task Force illustrated additional justification for the implementation of field solutions used to decrease fuel dependence. The study found that DOD operations suffered from degraded capability caused by unnecessarily high battle space fuel demand and fuel support costs, as well as unnecessary risk to support operations.[36] The task force asserted that implementing recommendations from its 2001 study would have increased capability and lowered fuel support costs.[37] In addition, the authors found that the DOD did not understand the direct relationship between its war fighting capabilities and fuel logistics.[38] To clarify, the DOD did not work on the observation that decreases in logistic requirements directly increases the tactical "tooth."[39]

The task force recommended two immediate ways to reduce the DOD's fuel dependence. One recommendation stated that there were multiple technologies available for all categories of

---

[34] Crowley, *Transforming the Way DOD Looks at Energy*, iii.

[35] Ibid.

[36] U.S. Office of the Under Secretary of Defense for Acquisition, Technology, and Logistics, *Defense Science Board Task Force Report on DOD Energy Strategy "More fight - Less fuel"* (Washington, D.C., 2008), 3, http://www.acq.osd.mil/dsb/reports/ADA477619.pdf (accessed August 24, 2010).

[37] Ibid., 23.

- The study directed the implementation of Fully Burdened Cost of Fuel, incentivized fuel efficiency programs throughout the DOD, and including fuel efficiency in the requirements and acquisition processes. Fully Burdened Cost of Fuel is the full burden of a gallon of fuel, which includes the purchase, transportation, protecting, storing, and delivering cost where the true cost hidden in the functions conducted and absorbed by the Services. This cost can easily range from $10 to $400 per gallon.

[38] Ibid., 18.

[39] Ibid., 23.

deployed systems at all levels of maturity that could reduce fuel demand.[40] An additional effort

recommended was the insulation of tents on forward operating bases to reduce air conditioning

and heating losses.[41] The DOD recently acted on the tent insulation recommendation as shown in

a 2010 Naval Post Graduate thesis.[42] The author of the thesis found that the implementation of

this technology reduced overall generator fuel consumption by nearly 50 percent.[43] This is one of

many examples of the DOD using the data from these studies to justify the start-up costs of

implementing fuel reduction technologies.

A Congressional Research Service (CRS) report summarized the DOD's fuel

consumption habits and illustrated the rising fuel costs over the past decade. Because of the

drastic cost increase, Congress directed that the DOD consider using alternate fuels to meet its

needs and to stimulate commercial interest in supplying those needs.[44] The report pointed out that

the issue for the DOD is not fuel usage but that costs have skyrocketed 373 percent.[45] Fuel

expenditures rose from $3B in 1997 to over $11B in 2007 while the consumption of fuel, even

with wars in Iraq and Afghanistan, has not increased significantly.[46] This study directly relates to

the hypotheses of this monograph and attempts to link moderate decreases in fuel consumption

and large decreases in fuel costs.

---

[40] U.S. Office of the Under Secretary of Defense AT&L, *"More fight - Less fuel,"* 23.

[41] Ibid., 33.

[42] Allen Rivera, "Cost Benefit Analysis of integrated COTS Energy-Related Technologies for Army's Force Provider Module" (master's thesis, Naval Postgraduate School, September 2009), 22, http://handle.dtic.mil/100.2/ADA509914 (accessed November 12, 2010).

[43] Ibid., 59.

[44] Congressional Research Service, *Department of Defense Fuel Spending, Supply, Acquisition, and Policy Report for Congress,* by Anthony Andrews, R40459 (Washington, DC: U.S. Library of Congress, 2009), 1, http://assets.opencrs.com/rpts/R40459_20090922.pdf (accessed October 15, 2010).

[45] Ibid., 2.

[46] Ibid., 3.

Tactical Solutions

Each of the previous studies and reports alluded to possible solutions for reducing fuel consumption. The remainder of this section discusses some of the literature illustrating those solutions in action. First, a National Defense article by Stew Magnuson asserted that tactical generators in Afghanistan and Iraq are the largest consumer of fuel on the battlefield.[47] Additionally, the author quoted a Government Accountability Office (GAO) study as stating that the DOD lacks an effective approach to reducing fuel demand, especially at remote outposts that rely on generators and are not on a local power grid.[48] According to an interview in the article, the program manager for mobile electric power stated that generator efficiency has not changed in over 40 years.[49] Magnuson presents two key points, which further support the purpose of this monograph. He first recommended the use of the new Advanced Medium Mobile Power Sources (AMMPS). These are new generators the Army is fielding, which have higher fuel efficiency ratings over older models. The Army intends to use these new systems to replace the current legacy generator inventory.[50] Second, he recommended the use of the Central Power Initiative (CPI). This is a program to replace multiple smaller generators with fewer, larger, more efficient generators, to manage power on outposts and decrease fuel consumption.[51] This recommendation also included the incorporation of a microgrid technology, the focus of one of the following case studies. Magnuson concluded that what drove the Army to treat fuel efficiency more seriously

---

[47] Stew Magnuson, "Ensuring No One Pays the 'Ultimate Price' for Fuel Becomes New Goal," *National Defense*, August 2009, 54.

[48] Ibid.

[49] Ibid.

[50] Ibid.

[51] Ibid.

was not only the economic cost, but also the cost of the soldiers killed when ambushed during convoy operations.[52]

In the LMI study mentioned previously, the authors conducted numerous surveys on crosscutting technology such as solar and wind, and garbage-to-electric generators, aimed at local electricity generation.[53] The demand for these technologies in theater stemmed from a joint urgent operational need statement in 2006 from Maj. Gen. Richard Zilmer, a Marine Corps Commander in Iraq's Anbar Province.[54] A quote from his request summed up the concerns with the operational risk of delivering fuel. General Zilmer asserted "by reducing the need for petroleum-based fuels at our outlying bases, we can decrease the frequency of logistics convoys on the road, thereby reducing the danger to our marines, soldiers, and sailors."[55] The services answered his operational need statement with the shipment of numerous commercially available, off-the-shelf technologies to his area of operations. His desire to force the fuel reduction issue, while tying his request to the reduction in the probability of the loss of life energized the defense community.

As illustrated in a 2001 study, logistic vehicles consume more fuel during wartime than combat vehicles.[56] In a top ten list of the highest fuel consuming vehicles, shown in Appendix 1, combat vehicles and rotary aircraft represented only four of the ten, where six logistic vehicles claimed the remaining top spots.[57] Because these systems use 1970s era technology for propulsion, the DOD realized an opportunity to inject new technologies on these platforms

---

[52] Magnuson, "Ensuring No One Pays the 'Ultimate Price' for Fuel Becomes New Goal."

[53] Crowley, *Transforming the Way DOD Looks at Energy*, E-24.

[54] Ibid., E-25.

[55] Ibid.

[56] U.S. Office of the Under Secretary of Defense for Acquisition, Technology, and Logistics, *Defense Science Board Report on More Capable Warfighting through Reduced Fuel Burden* (Washington D.C. 2001), 43, http://www.acq.osd.mil/dsb/reports/ADA392666.pdf (accessed August 24, 2010).

[57] Ibid., 43.

significantly decreasing fuel consumption during wartime.[58] One such technology, hybrid electric drives, continues to undergo testing and development within the defense community. However, hybrid electric vehicle technology, which began testing in the Army as early as World War II, was neither mature nor efficient enough to compete with mechanical systems until 1993.[59] As the technology matured, the Army identified hybrid electric vehicles as a potential technology that could meet the Army's future needs and provide expanded mission capabilities to warfighters.[60] The DOD now has the ability to test new hybrid technologies using the Hybrid-Electric Vehicle Experimentation and Assessment (HEVEA) program.[61] This program represents significant DOD investment in this emerging technology as well as expanding interest in hybrid electric vehicles.

The one unanimous theme from the literature is that policy finally created a culture within the defense community ready to address the cost crisis created from fuel use. The Army recognized the need to use technology to offset the exceedingly high and ever increasing cost of fuel and the burden that fuel places on the defense budget. However, what these sources do not take into account is the enterprise approach of using the Fully Burdened Cost of Fuel (FBCF) to estimate fuel cost.[62] Leaving this essential tool out of every fuel reduction calculation provides inaccurate data and demonstrates how fuel costs are historically underestimated. Additionally, the government does not have an enterprise approach to the implementation of fuel reduction technologies across the services. In the author's opinion, each service continues to have a stove-piped approach, which jeopardizes the widespread use of these technologies. Furthermore, the

---

[58] U.S. Office of the Under Secretary of Defense, *More Capable Warfighting through Reduced Fuel Burden*, 43.

[59] Ghassan Khalil, Major Christine E. Allen, and Michael Pozolo, "Hybrid-Electric Vehicle Experimentation and Assessment (HEVEA) Program Supports the Army's Need for Increased Power Demands," *Army AL&T Magazine*, January-March, 2009, 50-53.

[60] Ibid.

[61] Ibid.

[62] Fully burdened costs include standard fuel price, direct ground fuel infrastructure, indirect base infrastructure, environmental costs, delivery asset operations and support, delivery asset depreciation, and other specific costs. Source: Army Environmental Policy Institute, *Energy and Water Costing*, vii.

Army faces significant start-up costs when implementing new technologies. When attempting to offset these start-up costs, the combined services could benefit from an enterprise approach. Only the LMI report discussed this variable and the impact it has on budgets, which the Army forecasts in the long term. The link between the above national, operational, and tactical literature groupings show that decreasing fuel consumption at the operational level leads to decreases in both risk and dependence on foreign oil.

This monograph builds upon the policy recommendations by examining the efficiency of the recommended operational solutions to reduce the Army's fuel consumption. If the scholars and researchers are correct, reduced fuel expenditures resulting from reduced fuel consumption of fossil fuels requires immediate fielding of emerging fuel technologies.

Methodology

The primary goal of this monograph's analysis is to test the hypotheses that relate to the reduction of the Army's tactical fuel consumption. As such, this section presents the methodology employed to test the hypotheses in this monograph. This monograph first mentions the following proposed thesis in the introduction. *The use of emerging fuel technologies reduces the Army's fuel consumption and reduces deployed logistics requirements.* This monograph seeks to support this statement by using two hypotheses.

> *1. When there is an increase in the use of fuel-efficient technologies, there is a decrease in fuel consumption and a decrease in fuel costs occur.*

> *2. When there is an increase in the use of fuel-efficient technologies, there is a decrease in fuel consumption and a decrease in the risk to soldiers transporting fuel.*

This section consists of four subsections: selection of significant cases, instrumentation, data collection, and data analysis.

Selection of Significant Cases

As illustrated in the literature review, many technologies currently available could reduce the Army's deployed fuel consumption. However, this research focuses on the contemporary

16

environment for answers and aims at technologies, which can decrease the fuel consumption of the top ten battlefield users discussed earlier. Specifically, how does fuel efficiency increase and fuel consumption decrease with the use of three emerging technologies? The first emerging technology analyzed is the use of microgrids on forward operating bases. These microgrids reduce the number of small inefficient generators by combining and loading generators correctly, which make them operate more efficiently.[63] The second is a solar and wind power generation platform combining photovoltaic (solar) and wind power generation in order to augment or replace the use of inefficient generators.[64] The third is a diesel electric engine, or hybrid electric engine, which replaces legacy engines originally procured and installed without fuel efficiency as a consideration.[65] The next subsection establishes parameters to test these three case studies.

Instrumentation

Having selected three cases, which illustrate a fuel reduction capability, the next step requires a discussion on the instrumentation used to evaluate the fuel and budgetary savings of each technology. In order to test the hypotheses, the analysis of the cases must include questions aimed at generating data from the new emerging technologies in order to compare them to existing fuel consumption data. This monograph accomplishes this through four focused questions, listed below, and applied to each of the case studies.

How many units of the emerging technology are required to support the predetermined troop strength and/or the number of forward operation bases?

---

[63] See the following website for a concept of microgrids: http://www.lockheedmartin.com/products/intelligentmicrogridsolutions/MicrogridsVideo_01.html; or http://www.lockheedmartin.com/products/intelligentmicrogridsolutions/index.html

[64] See the following websites for brief descriptions on possible solar and wind emerging technologies: http://www.windtamerturbines.com/windtamer-turbines/; or http://www.skybuilt.com/products/ products_skystation.htm

[65] For further information about Oshkosh's HEMTT A3 Diesel-Electric see the following website: http://www.oshkoshdefense.com/products/12/hemtt-a3-diesel-electric

What is the overall fuel decrease following the incorporation of the emerging technology?

What is the overall decrease in fuel costs shown in dollars per gallon from the DLA-Energy (Defense Logistics Agency – Energy) price guide and dollars per gallon using the Fully Burdened Cost of Fuel (FBCF) analysis?[66]

What is the overall reduction in the number of 5,000-gallon tankers delivering fuel in one year?

Using these questions to focus analysis on each of the case studies makes possible the capturing of data from the theoretical implementation of each of the fuel saving technologies. This leads to extrapolating cost savings as well as determining the reduction of fuel consumed.

The first focused question tests the validity of using the emerging technology to replace or reduce the inefficient technologies. The concept is to devise a theoretical quantity of either microgrids, solar/wind generators, or hybrid electric powertrains required to support predetermined troop strength. The second question aims at capturing data illustrating the quantity of fuel saved by using the new technology. This data provides additional information for the third question; what is the decrease in the cost of fuel over a year, and the fourth question; what is the reduction of the quantity of tankers required to deliver the fuel. Answering both the third and fourth focused questions either support or undermines the hypotheses in this monograph. This point in the monograph is an appropriate time to add to the compelling argument for the use of these technologies. Using the data for the number of tankers removed from the road allows the calculation of the number of casualties avoided. Regardless of the value found here, it is important to note that implementing these technologies could have increased or decreased casualty vales depending on the environment and the threat. However, the possibility of a theater

---

[66] Fully burdened costs include standard fuel price, direct ground fuel infrastructure, indirect base infrastructure, environmental costs, delivery asset operations and support, delivery asset depreciation, and other specific costs. Disagreement exists on the application of delivery asset depreciation due to the number of delivery assets not being directly scalable to fuel consumption. Source: Army Environmental Policy Institute, *Energy and Water Costing Methodology.*

aggregate decrease in deaths is a compelling argument for further research of the implications of fuel saving technology implementation.

Data Collection

For this monograph, it is necessary to collect data pertaining to the three emerging technologies as well as data used to evaluate their usefulness as a fuel-efficient technology. Historical data, professional journals, government agencies, and energy sector databases provide the necessary parameter data for each of the case studies. The combination of existing studies from RAND, the Government Accountability Office, and LMI combined with technology fact sheets from the Office of Naval Research and the Marine Corps Expeditionary Office provide the necessary background performance and cost parameters for the technologies. Using the data found throughout each of these sources produces sufficient information to select a specific microgrid system, solar/wind power generation package, and a hybrid electric powertrain on which to conduct an analysis.

In order to conduct the same analysis on each case study using the focused questions and generate useful data, this study requires five control variables. These control variables establish the limits of comparison of the old technologies against the emerging ones. Limiting these variables to a specific region, the Iraqi Theater of operations in FY2007, and on a specific level of government, DOD or Army, allows for simple data collection. These variables still answer the critical focused questions on fuel savings while delimiting the continental U.S. and other overseas locations, which are outside the scope of this monograph. The first control variable, troop strength, defines the number of soldiers in an expeditionary location that the emerging technology supports. Using 165,000 as the number of soldiers deployed to Iraq in FY2007 establishes a baseline for applying the new technologies.[67] Using primary sources from the DOD, the

Brookings Institute produces a federal government accepted standard of troop strength per year in

its annual Iraq Index. The second control variable is the number of bases in Iraq in FY2007. Due

to the sensitive nature, high turnover of locations, and rapidly changing troop strength, using an

average of 100 bases is an acceptable value.[68] The third control variable is the number of gallons

of fuel transported in one year. This establishes a number by which the study can determine the

decreased risk to logistic soldiers as well as aid in determining a decrease in the Fully Burdened

Cost of Fuel (FBCF). The Army Environmental Policy Institute (AEPI) Report recognizes this

value as 502 million gallons transported in FY2007.[69] The fourth control variable is the average

fuel consumption per year for the expeditionary force. Since no agencies tracked fuel consumed

by type; JP-8, or diesel, in the Iraqi Theater in FY2007, this monograph assumes that the amount

of fuel purchased in the theater equals the amount of fuel consumed. Using a value of 18.9

million barrels, or 793.8 million gallons of fuel purchased as the fuel consumed for the Iraq

Theater in FY2007, allows this study to illustrate the impacts of fuel reduction using emerging

technologies.[70] The fifth control variable is the average fuel cost per year. DLA-Energy provides

---

[67] Brookings Institute, *Iraq Index: Tracking Variables of Reconstruction & Security in Post-Saddam Iraq*, by Michael E. O'Hanlon and Ian Livingston, December 9, 2010, 19, http://www.brookings.edu/iraqindex (accessed 20 December 2010).

- In order to calculate the number of soldiers in Iraq in FY2007, the author used an average from the *Iraq Index: Tracking Variables of Reconstruction & Security in Post-Saddam Iraq* which showed the following values for deployed soldiers. June 2007 - 150,336; January 2008 – 155,846; June 2008 – 182,060.

[68] U.S. Department of Defense, Office of the Assistant Secretary of Defense (Public Affairs) "DoD Announces Units for Next Operation Iraqi Freedom Rotation," June 27, 2006, http://www.defense.gov/releases/release.aspx?releaseid=9770 (accessed December 30, 2010).

- Global Security, "Iraq Facilities," http://www.globalsecurity.org/military/facility/iraq.htm (accessed December 21, 2010).

Using DOD News releases notifying units of deployment with a Global Security website illustration on the locations of forward operating bases allows the estimation of 100 bases in Iraq in FY2007.

[69] Army Environmental Policy Institute, *Sustain the Mission Project: Casualty Factors*, 5.

[70] Konrad Stutzmann, Defense Logistics Agency – Energy, e-mail to author, December 24, 2010.

a well-established cost per gallon but does not calculate the Fully Burdened Cost of Fuel (FBCF). Additionally, DLA-Energy does not provide data for fuel cost in a specific region. According to the Congressional Research Service Report to Congress, the DOD spent more than $2.98 billion in FY2007 on the purchase of fuels in the CENTCOM Area of Responsibility (AOR).[71] Dividing the cost of fuel purchased for the CENTCOM AOR by the amount of fuel purchased gives a price per gallon for the CENTCOM AOR of $2.21. DLA-Energy bases the cost per gallon on the yearly purchase of bulk fuel based on the market price. However, what the cost per gallon does not include is the transparent costs that the FBCF captures. Using the AEPI Reports' value of FBCF for a mature theater of $14.13 per gallon and the amount of fuel purchased for the CENTCOM AOR to calculate the total cost of fuel this study arrives at $19.08 billion for FY2007.[72] The data collected above allows straightforward illustration of the impacts the three emerging technologies have on fuel consumption and cost.

Data Analysis

This subsection describes how to generate the data required in this comparative analysis and the steps taken to map the study of each case. Comparing the emerging fuel technologies to existing ones, while holding the control variables constant, generates easily interpreted raw data. After applying the research questions to each case study, the data should lead to the conclusion that increases in the use of new fuel-efficient technologies, decreases the budgetary outlays for fuel and decreases risk to soldiers, proving both hypotheses.

The first step in conducting an analysis on the three cases is to apply both the first and second focused research questions, generate numbers for the quantity of equipment needed, and

---

[71] Congressional Research Service, *Department of Defense Fuel Costs in Iraq,* by Anthony Andrews and Moshe Schwartz, R40459 (Washington, DC: U.S. Library of Congress, 2009), CRS-4, http://www.policyarchive.org/handle/10207/bitstreams/18707.pdf (accessed December 20, 2010).

[72] Army Environmental Policy Institute, *Energy and Water Costing Methodology,* vii.

the gallons of fuel saved. In order to compute the number of units required, the analysis substitutes the existing technology in the control group with the new technology or replaces the need for the existing technology completely. Additionally, in order to compute the theoretical fuel consumption savings, the analysis either uses industry produced values on fuel savings or the overall reduction of fuel consumption due to removal of the existing technology.

The second step is to use the data from questions one and two and compute the decrease in fuel cost in dollars per gallon for each of the technologies. Computing two values of fuel savings, one based on DLA-Energy's cost and one on the FBCF analysis, answers the question of the savings per year from implementing the new technology. These values illustrate how fuel consumption affects the DOD and Army's annual budget.

The third step is to compute the reduction in the number of 5,000-gallon tankers removed from the roads in one year due to the decrease in fuel requiring transport. This value comes from subtracting the overall fuel decrease in a year in step one from the total fuel transported in Iraq in FY2007. Knowing the number of trucks no longer needed to convoy fuel leads to the calculation of the number soldier deaths avoided by using the emerging technology. Using the AEPI study's casualty factor enables this monograph to analyze the link between the decreases in fuel required to the decrease in the probability of a soldier killed convoying fuel. The decreased demand for transport trucks, logistic personnel, and fuel to transport fuel makes a clear argument for including the causality factor in the FBCF in order to property project Army fuel budgets.

Summary

This section first restated the research question and the hypotheses. It then attempted to illustrate the rational for choosing the three case studies as well as briefly describing each. Following this section's introduction and case study description was a clear definition of the focused questions applied to the case studies aimed at proving the hypotheses. Next was a subsection with data collection methods for each case study and control variables with justification for including them in this monograph. Concluding this section was a complete

systematic description on how to conduct this study at a later time. Each of the above-mentioned

subsections lays the foundation for the following section, the analysis of each case.

Analysis of Emerging Fuel Technologies

Microgrids

Serious deficiencies in the Army's ability to match power demand to generation arose

during Operations Iraqi and Enduring Freedom (OIF and OEF). The shortfall became apparent

when the force became dependent on an excessive quantity of civilian generators. These systems

met the immediate need; however, they supplied power on an individual basis. In addition, the

ability to use military consumption factors to track fuel use disappeared. The Army began to build

operating bases with the creature comforts reminiscent of home. Power demand exceeded

military generators' capacity and with a mix of military and civilian power generator platforms,

fuel consumption spiraled out of control. An alternative solution to meeting the demand would be

to manage the power requirements just as the civilian sector does, using a grid. A power grid

allows power generation facilities to go on or offline dependent on demand. This management

technique conserves resources and matches supply to demand. Allowing the Army to manage

power on operating bases in the same manner would reduce the need to for such an excessive

quantity of fuel to power these generators.

After realizing the excessive fuel consumption and the inordinate amount of generators

required to support operations in Iraq and Afghanistan, the Army began looking toward the

private sector for new technologies to manage power distribution and loads on generators. One

new technology, called microgrids, allows for a more optimized loading of the generators so that each can operate within an efficient range reducing fuel consumption.[73] For example, if there were five generators on a forward operating base and each averaged less than a 30 percent load for a 24 hour period, a system that could combine these generators and turn them on an off based on demand has the capability to drastically reduce fuel consumption. This is the concept behind microgrids.

A microgrid can be in many configurations, but for the Army's use, it includes three primary devices and a host of ancillary equipment.[74] The first device, the power management controller, is a mirror image of how power transmission companies in the U.S. controls the civilian grid where, based on demand, power sources are included and excluded. One exception is that an Army microgrid provides the capability to shut down and startup numerous power generators in order to match supply to demand and not vice versa. The second key component, a power-conditioning controller receives electricity from the array of power generators and conditions the power. This is imperative due to the restraints of combining electricity from differentiating power sources where frequencies and voltages do not match. The third device is a suite of feedback and distribution boxes used to provide data to the controls as well as distribute power to the loads on the microgrid. This device is unlike the power distribution panels in the Army's inventory in that the new panels provide feedback and communicate with the controllers. The Army chose to develop this technology so that the DOD might reap the benefits of a system operating multiple small generators as if they were a larger, more efficient, and correctly loaded one.

The purpose of a microgrid is to combine power generation from multiple sources and condition it in a way as to produce clean, uninterrupted power. The reason this technology

---

[73] See the following website for a concept of microgrids: http://www.lockheedmartin.com /products/intelligentmicrogridsolutions/index.html (accessed February 15, 2011)

[74] Teri Hall, Lockheed Martin, "Microgrid Development for Tactical Operations" (presentation at the 2009 Joint Service Power Expo, New Orleans, LA May 5, 2009).

interests the Army is that the technology allows the input of multiple hybrid and conventional power sources and runs each of them at the optimal level on a grid.[75] With generators being the number one fuel consumer in Iraq and Afghanistan in either a mature or an immature theater, the Army wants to use this technology to reduce fuel consumption. A microgrid also generates the capability to join power sources now and in the future. For instance, the opportunity exists today to use this technology to join older, legacy generators with future power sources such as photovoltaic, wind, and even garbage-to-energy generators. This technology fits well with the Army's next generation of generators called the Advanced Medium Mobile Power Source (AMMPS) which has the capability to run at variable speeds based on load and can be powered down when connected to and controlled by a microgrid.[76] The ability of the AMMPS to power down presents significant fuel savings over current models of Army generators. The research and development of microgrid technology has national level relevance since it demonstrates an additional technology the DOD intends to use to reduce overall fuel consumption and in turn decrease dependence on oil.

In a telephone interview with the author, Paul Richard revealed that in past analysis of generator usage, most generators only see a ten to thirty percent load during operations.[77] This means that the majority of generators operates extremely inefficiently and risk wet stacking.[78] Inefficient operation also leads to excessive fuel consumption illustrated in the current fleet of generators, which consume nearly the same amount of fuel under load or at no-load conditions. In

---

[75] Dina Fine-Maron, "Can a Microgrid Protect U.S. Troops in Afghanistan?" *Scientific American*, December 22, 2010, http://www.scientificamerican.com/article.cfm?id=can-a-microgrid-protect (accessed February 10, 2011).

[76] Paul Richard, interview by author, February 15, 2011. Paul Richard is the Project Manager for Mobile Electric Power. His office controls the Army and Air Force life cycle management for generators. Source: Project Manager Mobile Electric Power website: www.pm-mep.army.mil/

[77] Ibid.

[78] Wet stacking refers to a generator running at a high rpm and experiencing little to no load. This causes maintenance issue like excessive oil blow-by and bearing and seal damage. Generators operating under ideal conductions experience a 70-90 percent electrical load.

order to decrease these fuel costs, a microgrid keeps loads in the most efficient range between seventy and ninety percent.

The Army could realize a thirty percent fuel savings if it implements this technology on operating bases with three to ten power sources and no civilian grid.[79] This savings does not mean that every generator connected to the microgrid experiences this level of savings; the savings is a combined fuel savings of all the systems on the microgrid. Additionally, the opportunity exists to reduce long-term maintenance and repair costs that fall outside the scope of this monograph. Operationally, this technology has the potential to address excessive fuel consumption in multiple expeditionary locations exemplified by the study of fuel use at Contingency Operating Base (COB) Adder in 2008.[80] This COB consumed more than 1.6 million gallons of fuel in June of 2008 where 1.17 million gallons of fuel, or seventy percent, sustained base support activities.[81] This COB is an excellent example of the multiple locations throughout Iraq where implementing this new technology could contribute to a theater wide thirty percent fuel savings.

The Army intends for this technology to replace redundant, medium power generators operating independently. Since combining low power, less than 100 kilowatt (kW), generators onto a grid on bases is a new concept, the Army benefits by spreading the electrical load across less generators, which allows more efficient operation. However, this suite of devices increases the upfront logistics requirements across the services. Additionally, any new analysis of the Fully Burdened Cost of Fuel (FBCF) calculations would need to include this technology. This ensures future studies capture the difference between fuel consumed in transporting this technology and fuel saved by implementing this technology.

---

[79] Hall, "Microgrid Development for Tactical Operations."

[80] U.S. GAO, *Defense Management: DOD Needs to Increase Attention*, 46.

[81] Ibid.

Since this technology is still in the development phase, this monograph makes two assumptions in order to compute the fuel savings while addressing how the technology applies to the hypotheses. Current designs for microgrids aim at joining the power production of six or more power generation sources. However, the data available from the research and development studies conducted on this new technology only illustrates the fuel savings of combining three generators.[82] In order to calculate fuel savings, this case study assumes that a microgrid can handle six power inputs. Additionally, in order to create patterns of use on operating bases, this case study uses the COB Adder example from above to estimate the number of generators on a operating base as well as the total fuel consumed by a base not on a civilian power grid. The assumption now is that all units located on the operating bases could take advantage of this emerging technology. Therefore, the number of military generators currently in Iraq equals the number of generators on all operating bases in FY2007. This enables the calculation of the fuel savings from implementing microgrid.

In order to test this monograph's hypotheses, these next paragraphs address each of the four focused questions. The first question is how many microgrids are required to support the predetermined troop strength and the number of forward operating bases? With over 100 operating bases in Iraq in FY2007 varying in size from small combat outpost to large bases such as Joint Base Balad, calculating an acceptable number of microgrids must come from an accurate database. According to an email from the office of the DA G4 who manages the database of all equipment currently in Iraq, there were 1,505 generators remaining and undergoing processing for retrograde.[83] Understanding that this is a extremely low value due to the retrograde operations ongoing in Iraq and the inability to capture all the civilian and military generators in theater at any time, leads to an approximation of at least fifteen generators per operating base. Assuming

---

[82] Hall, "Microgrid Development for Tactical Operations".

[83] Gregory Bowie, e-mail message to author, February 10, 2011.

there were at least this many generators in FY2007, this leads to 250 microgrids. This value allocates at least two microgrids per base with a fuel savings of thirty percent for each.

The second focused question is what is the overall fuel decrease following the incorporation of the microgrids? Using an average fuel consumption factor of 0.83 gallons per hour for each of the 1,505 generators located on the forward operating bases leads to a yearly consumption of 10.9 million gallons per year.[84] With the incorporation of the microgrids and the corresponding thirty percent fuel savings, this could lead to a 3.3 million gallon per year fuel savings.

The third focused question is what is the overall decrease in fuel cost? The data calculated from this question presents itself in two ways: dollars per gallon from the DLA-Energy price guide, and dollars per gallon using the Fully Burdened Cost of Fuel (FBCF) analysis. Using the 2007 values for each, the first is DLA-Energy's cost of $2.21 per gallon and the second is the FBCF value of $14.13 per gallon.[85] Basing calculations for the decrease in cost on these values leads to a DLA-Energy fuel cost savings of $7.3 million and FBCF cost savings of $46.6 million. Since the FY2007 budget for operations in OIF was $131.2 billion, these cost savings account for less than a tenth of a percent and a 0.04 percent reduction in the DOD's FY2007 budget.

The fourth focused question is what is the reduction in the number of 5000-gallon tankers no longer needed to deliver fuel? The data calculated from this question produces a value for the reduction for an entire year. As stated previously, this study uses the Army's standard fuel transport truck to illustrate the impact on the use of Army soldiers, not locally contracted transport. The amount of fuel saved after linking the generators to the microgrids reduces the number of tankers conducting convoys by 660 per year. According to the casualty factor study,

---

[84] In order to calculate a consumption factor of 0.83 gallons per hour for the five kW (0.57 gallons per hour) and ten kW (1.09 gallons per hour) generators in Iraq, the author took the average of two generators combined assuming an even distribution of each in theater.

[85] Army Environmental Policy Institute, *Energy and Water Costing Methodology*, vii.

the reduction of 660 tankers per year leads to the reduction of the probability of the loss of life. One soldier may not have died delivering fuel in 2007.[86] This value sufficiently illustrates the risk reduction to soldiers, which partially supports this monograph's hypotheses.

As this study demonstrates in the calculations, unless the number of microgrids is significantly higher, the DOD budget savings is not the driving factor in implementing this technology. Unfortunately, the above calculations for the fuel savings of microgrids should not be the basis on the Army's decision to invest in this technology. However, if this study used COB Adder as the standard for fuel used by generators, the amount of fuel saved and the number of tankers decreased significantly changes. For instance, as stated above, fuel consumption on COB Adder was 1.17 million gallons in one month. If this study makes a simple assumption, that seventy percent of 1.17 million gallons was specifically for power generation (not an unreasonable assumption in June for a sprawling thirty square kilometer base) then in one month fuel consumption by the generators exceeds 819,000 gallons. Using microgrids on COB Adder leads to a thirty percent savings of 245,000 gallons, which is forty-nine 5,000-gallon tankers, or three sixteen-tanker convoys each month. This example obviously cannot be the standard for all the operating bases in Iraq due to the size of Adder. However, even the use of microgrids in niche applications has the potential for significant reductions in fuel consumption and risk.

Implementing this technology in the Army would increase the logistics required to transport and maintain this new equipment, but these additional requirements are well worth the cost. However, unless the operating base has more than three generators, the technology would be prohibitive since it would require one transport truck to deliver the microgrid components and potentially only remove one 5000-gallon tanker from the road in one year. As such, this study advises reserving microgrids for applications where numerous generators operate, unlike small combat outposts.

---

[86] Army Environmental Policy Institute, *Sustain the Mission Project: Casualty Factors*, 6.

Furthermore, the DOD should embrace this technology, assume the startup costs, and spread the cost over the lifetime of the product against the national security benefits. These benefits could include a small budget savings but a significant decrease in the cost in lives and operational risk to each mission. Additionally, what this means at the national level is that there would be a significant decrease in the amount of fuel convoys delivering to multiple fuel storage locations throughout Iraq. This is particularly useful in Afghanistan where the military does not control fuel convoys until they get through Pakistan. There is a significant risk in public opinion and risk caused by the lack of a sufficient fuel supply. Commanders in Afghanistan face significant operational risk while the nation faces high fuel delivery costs if this fuel supply were interrupted for a lengthy period of time.

Solar and Wind Power Generator

Relying solely on generators to produce mission essential power at austere operating bases in Iraq in 2006 illustrated an operational vulnerability. Concern throughout the DOD arose when Maj. Gen. Richard Zilmer requested readily available commercial technology to help him overcome his dependence on the vulnerable fuel convoys.[87] The immediate response to Maj. Gen. Zilmer's request by the services was to send commercial versions of solar and wind power generation platforms to Western Iraq. Seven years later the DOD is only one-step closer with the testing of solar and wind equipment to devising an alternative to the internal combustion engine generator, which the Army needs to sever the fuel umbilical cord to remote sites. The Army is attempting to test and develop a combined solar and wind power generation station, the focus of this case study, in an attempt to lower Iraq and Afghanistan's theater fuel consumption.[88] The

---

[87] Crowley, *Transforming the Way DOD Looks at Energy*, E-25.

[88] See the following websites for brief descriptions on possible solar and wind emerging technologies: http://www.windtamerturbines.com/windtamer-turbines/; http://www.skybuilt.com /products/ products_skystation.htm (accessed February 15, 2011).

intent of this case study is to analyze the fuel savings of this system and identify if solar and wind power generation is a suitable alternative to the Army's Tactical Quiet Generator (TQG) sets.

A solar and wind power generator is a wind-turbine coupled to photovoltaic (solar) panels. The technology in the Army must meet strict reliability requirements in order to provide uninterrupted power to tactical operations centers and essential mission support activities at remote locations. Any solar and wind power generator fielded to the Army would include either a container or trailer for deployability, a bank of batteries, a wind turbine, solar panels, power inputs, and the power conditioning electronics. There exists the capability for this technology to produce moderate amounts of power; however, the Army requires that this be a small system capable of producing five to ten kilowatt-hours (kWh) of power. The system would produce voltages suitable for augmenting military vehicle power and providing power for computers and lights on a small scale.

According to the Project Manager for Mobile Electric Power, the Army strategy for these systems is to meet niche operational capabilities for reliable remote power. This minimizes logistical and operational impacts and most importantly, reduces the fuel consumption related to power generation.[89] More specifically, the Army's intent for solar and wind technology is to produce power and consume no fuel. The system's design allows power production for austere operating bases with small power requirements such as a company or platoon tactical operations center with fuel delivery restrictions. The Army views this as an opportunity to eliminate the need for a fuel consuming internal combustion engine generator, which decreases the need of delivering fuel to these austere locations. Additionally, since the power output is small, the Army sees this as only a solution for small power requirements. Each application of this system could

---

[89] Paul Richard, Department of Defense Project Manager Mobile Electric Power, "Mobile Electric Power for Today and Tomorrow" (presentation at the 2009 Joint Service Power Expo, New Orleans, LA, May 5, 2009).

31

support a small number of laptops, lights, and radios as the solar and wind power generators would be incapable of powering a large environmental control unit.

Most solar and wind power generators supply stand-alone power dependent on location and conditions. If this technology replaced all of the existing internal combustion engine generators operating on austere bases between five and ten kW, then the Army would benefit from a fuel savings of 100 percent per system. This leads to a 5,000 gallon per year savings for five kW and 9,500 gallons per year for ten kW generators if these systems ran for twenty-four hours a day. For every generator at a forward operating base replaced by this system the Army also benefits from a direct savings in the delivery and storage cost of the fuel.

Due to the size limitations for solar panels and wind turbines stored and deployed out of a shipping container or trailer, the system could only be expected to produce up to ten kW of power. Because of this, the Army aims at using this technology to augment existing medium sized power generation platforms. Additionally, these systems would be unit specific and the Army would not field these to combat service support units. These units normally operate on enduring bases, as such as Joint Base Balad, with grid power and therefore would not be applicable to this study.

This study makes two assumptions in order to illustrate the possible fuel savings of the solar and wind power generation platforms. First, that the DOD could allocate the funding required to purchase solar and wind power generators while still fielding new internal combustion engine generators. The photovoltaic panels used in solar arrays are only five to twenty percent efficient and very costly. This startup cost would otherwise prohibit the procurement of this emerging technology. The second assumption is that the Army would procure this system even with the large diameter wind turbine required to generate the promised power. If these systems required the coupling of a two-meter square solar array with a wind turbine, then the diameter of the wind turbine would need to be eight feet. This width approaches the maximum dimensions of standard Army trailers and any wider turbine prohibits the use of the technology due deployment restrictions of large equipment.[90]

There are two limitations of a solar and wind power generation system: the ability to produce uninterrupted power and the low kW power producing capability. First, these systems require a location with sustained winds and exposure to direct sunlight to produce power in the five to ten kW range. The battery backups designed into each of these systems provide power during low light and wind conditions as well as during peak demand. Second, power output of these systems is proportional to their footprint. Larger power outputs, above ten kW, require more square footage of ground for the solar arrays and a larger diameter wind turbine. Once the power demands exceeded this systems power producing capability, the platform would need generator augmentation, further complicating the logistical support planning. The Army and DOD evidently realize these restrictions and still intend to find applications where this technology can significantly reduce dependence on fuel.

In order to support this monograph's hypotheses, these next paragraphs address each of the four focused questions. The first question is how many solar and wind power generators are required to support the predetermined troop strength and the number of forward operating bases? According to the Project Manager for Mobile Electric Power, no data exists which captured the number of low power, five and ten KW, military and civilian generators operating Iraq in FY2007.[91] However, based on the control variable for the number of soldiers in theater and the Modification Table of Organization and Equipment (MTOE) for the units notified for deployment to Iraq in FY2007, this study arrives at an estimate of 567 systems. [92] The generators, which

---

[90] Headquarters, Department of the Army and Air Force, TM-9-2330-394-13-P M1082 Series 2 ½ Ton Light Tactical Vehicle Trailer (LMTVT) and M1095 Series 5 Ton Medium Tactical Vehicle Trailer (MTVT) Manual, (Washington, D.C. Government Printing Office, January 2005), 00-07.

[91] Paul Richard, e-mail to author, January 28, 2011.

[92] U.S. DOD, "DoD Announces Units for Next Operation Iraqi Freedom Rotation."

belong only to the brigade combat teams who deployed in FY2007, range from five to ten kW. Additional fifteen and thirty kW systems are also organic to these units, but these occur in small quantities and are outside the scope of this study. The solar and wind power generators would only replace the tactical quiet generator sets specifically assigned to brigade combat teams and not the generators belonging to the combat service support units. The number of solar and wind systems required is a reasonable assumption based on the brigade combat teams operating at locations where there would be no electrical power grid. Additionally, this case recognizes that each unit's list of equipment authorized for deployment would be dissimilar. Some units would fall in on stay behind equipment and others would deploy their entire MTOE. Neither of these facts affects the overall number of systems required since the units would use generators left behind by redeployed units in Iraq or deploy with their organic systems in order to meet their power requirements.

The second focused question is what is the overall fuel decrease following the incorporation the solar and wind technology? Both five and ten kW generators consume fuel at a rate of 0.57 and 1.09 gallons per hour respectively.[93] Assuming the combined twenty-four hour usage for the length of a twelve-month deployment, these generators consume 4.6 million gallons of fuel in one year. As discussed above, the Army units in Iraq purchased and consumed 793.8 million gallons of fuel.[94] Therefore, with the incorporation of the solar and wind technology the Army could recognize a fuel savings of 4.6 million gallons, or a 0.58 percent reduction in fuel.

The third focused question is what is the overall decrease in fuel cost? The data calculated from this question presents itself in two ways: dollars per gallon from the DLA-Energy

---

- Based on the number of five kW and ten kW generators in a brigade combat team calculated from an MTOE analysis on the Force Management System (FMS) website for IBCT, HBCT, CAV and SBCTs. https://fmsweb.army.mil/unprotected/splash/ (accessed on January 30, 2011).

[93] U.S. Department of Defense, Standard Family of Mobile Electric Power Generation Sources General Description Information and Characteristics Data Sheets Handbook, MIL-HDBK-633A, January 26, 2010, http://www.everyspec.com (accessed on January 30, 2011).

[94] Konrad Stutzmann, Defense Logistics Agency – Energy, e-mail to author, December 24, 2010.

price guide and dollars per gallon using the Fully Burdened Cost of Fuel (FBCF) analysis. Using the FY2007 value for each, the first is the DLA-Energy's cost of $2.21 per gallon and the second is the FBCF of $14.13 per gallon.[95] Basing calculations of the decrease in cost on these values leads to a DLA-Energy fuel cost savings of $10.2 million and a FBCF cost savings of $65.1 million. Since the FY2007 budget for operations in OIF was $131.2 billion, these cost savings account for a 0.01 and a 0.05 percent reduction in the DOD's budget.

The fourth focused question is what is the reduction in the number of 5000-gallon tankers no longer needed to deliver fuel? The data calculated from this question produces a value for the reduction for an entire year. Using the amount of fuel saved from incorporating the solar and wind technology reduces the number of tankers conducting convoys by 921 per year. For every 5,000-gallon tanker removed from the road, one tractor and the associated logistics soldier driving it does not incur the operational risk and the Army benefits from a decreased chance of casualties. This study acknowledges that this value assumes only one truck and tanker would be needed for a point-to-point delivery of fuel. For example, a tanker leaving Kuwait with 5,000-gallons of fuel does not deliver fuel to each of the 567 generator locations. There are additional transportation requirements within theater further increasing the use of fuel to deliver fuel. Additionally, according to the casualty factor study, the reduction of 567 tankers per year leads to the reduction of the probability of the loss of life. One soldier may not have died delivering fuel in 2007.[96] This value sufficiently illustrates the risk reduction to soldiers linked with implementing this technology partially proving this monograph's hypotheses.

At the tactical level, this solar and wind power generation capability meets the demands of company and platoon tactical operations center in areas where there is no electric power grid.

---

[95] Army Environmental Policy Institute, *Energy and Water Costing Methodology*, vii.

[96] Army Environmental Policy Institute, *Sustain the Mission Project: Casualty Factors*, 6.

The system also alleviates the need for fuel, decreasing the Army's cost while also decreasing the risk to soldiers who transport fuel. However, the research in this case study leads to one concern about implementing this technology: the ability to produce uninterrupted power. This system depends on a constant wind source and direct sunlight or it suffers from a reduced electrical power output. This leads to commanders weighing their operational risk to their mission against the dependability of these new systems.

Operationally, this technology presents a viable alternative to power generation on austere operating bases. The Army benefits from decreased fuel consumption, increased availably of logistic soldiers, and decreased risk to logistic and force protection soldiers conducting fuel convoys. However, as illustrated by the above calculations, a $65.1 million decrease per year of operation would most likely not be the primary reason the Army invests in this emerging technology. What is apparent is the significant decrease of tankers not needed to deliver fuel. This fact has several second and third order affects providing compelling reasons why the Army should combine multiple technologies in an enterprise approach to providing power. The Army must find and use an emerging solar and wind technology in order to decrease the human cost of resupplying Army units.

HEMTT A3 Diesel Electric

In 2006, the Army spent $3.5 billion on facility energy used to power installations throughout the world, excluding deployed locations. Contrasting to that $3.5 billion, which seems to be a large sum, is the $16.5 billion spent during the same period to fuel tactical vehicles.[97] Furthermore, the Army owns 260,000 tactical wheeled vehicles as of November 2010, which presents opportunities to increase fuel efficiency and drastically decrease the national dependence

---

[97] U.S. Department of Defense, *Annual Energy Management Report: Fiscal Year 2006,* (Washington D.C: January 2007), 3, http://www.acq.osd.mil/ie/energy/energymgmt_report/ fy06/FY_2006_Annual_Report_ Narrative.pdf (accessed January 10, 2011).

on fuel.[98] In 2008, a study comparing tactical and combat vehicles found that wartime fuel consumption in the former exceeds that of the latter. The tactical vehicle fleet, comprised of mostly logistic vehicles, consumed over 173 million gallons of fuel.[99] Even a discernible improvement of ten to twenty percent in fuel efficiency could lead an increased fuel savings of twenty million gallons or more per year. This fact leads to the third technology capable of lowering fuel consumption: hybrid electric powertrains.

As demonstrated by the increased efficiency and the resulting high sales in hybrid cars in the civilian market, the fuel savings from using these powertrains could warrant the Army's research, development, and start-up costs. The application of this technology in the Army's systems could have significant impacts on fuel savings when comparing the size of the wheeled vehicle fleet with its excessive level of consumption. The Army aims to implement hybrid electric technology, which mirrors the systems found in civilian hybrid vehicles. However, Army versions must meet the same extreme performance capability standards as the existing tactical fleet. Additionally, the Army has not yet incorporated hybrid electric technology in any fleet wide application except for non-tactical vehicles. However, there are numerous studies underway to test the viability of the widespread use of this technology in wheeled systems. Recently, Congress bolstered interest by funding a feasibility and advisability study focusing on the possibility of using hybrid electric powertrains as part of the tactical wheeled vehicles program.[100] The Congressionally funded study, which began in 2003, combined the efforts of the product manager for tactical wheeled vehicles and the Oshkosh Corporation.[101] The intent was for the product

---

[98] U.S. Department of the Army, *Army Truck Program: Tactical Wheeled Vehicle Acquisition Strategy Report to the Congress,* (Washington D.C.: June 2010), 5, www.defenseindustry.com/2010_ US_Army_Tactical_Wheeled_Vehicle_Acquisition_Strategy.pdf (accessed February 15, 2011).

[99] U.S. Office of the Under Secretary of Defense AT&L, *"More fight - Less fuel",* 44.

[100] Mitchell J. Kozera, Product Manager Medium Tactical Vehicles Engineering Report to Congress, *Hybrid Electric Powertrains for Tactical Wheeled Vehicles,* e-mail to author January 13, 2011.

[101] Ibid.

manager to guide Oshkosh's research and development efforts to produce a hybrid electric version of the Heavy Expanded Mobility Tactical Truck (HEMTT), a family of vehicles currently used throughout the Army.[102] This case study analyzes the HEMTT A3, a hybrid electric version.[103]

There are numerous types of hybrid electric systems, but three designs give a sufficient description: mild, parallel, and series hybrids.[104] The technology in the HEMTT A3 is a diesel electric *series* hybrid, which uses a single path to power the wheels, but two energy sources. The internal combustion engine turns a generator, which charges the batteries or drives electric motors powering wheels. The generator can also charge the batteries during low load conditions. The benefit of a series hybrid over mild or parallel is that the internal combustion engine can be smaller and more efficient since its only use is to increase the battery voltage and not drive the wheels. The Army chose this technology because hybrid electric powertrains are a proven, dependable alternative when combined with an internal combustion engine. Additionally, the

---

[102] HEMTTs are a family of 10-ton, eight-wheel-drive vehicles designed to provide heavy transport capabilities for the supply of Army combat vehicles and weapons systems. The Army uses these systems throughout the logistics community specifically for cargo carrying and fuel delivery in brigade combat teams as well as support units. The A4 models, the most recent variant of the HEMTT, includes a cargo variant, 2,500-gallon refueler, fifth-wheel tractor, a load handling system (LHS) variant, a recovery vehicle with cranes and winches, a light equipment transporter and a guided missile transporter. Source: Oshkosh Corporation, *HEMTT A4 Information packet*, http://www.oshkoshdefense.com/products/13/hemtt-a4 (accessed February, 12 2011).

[103] For further information about Oshkosh's HEMTT A3 Diesel-Electric see the following website: http://www.oshkoshdefense.com/products/12/hemtt-a3-diesel-electric (accessed February 15, 2011).

[104] Mild hybrid uses an Internal Combustion Engine (ICE) and a integrated starter generator to augment the ICE driving the wheels. A parallel hybrid uses two paths to power the wheels, a conventional ICE path and a electrical path combined through a hybrid transmission. Source: Mitchell J. Kozera, Product Manager Medium Tactical Vehicles Engineering Report to Congress, *Hybrid Electric Powertrains for Tactical Wheeled Vehicles*, e-mail to author January 13, 2011.

hybrid electric systems have undergone extensive capability and endurance testing since the production of the first technology demonstrator version for the Army in 2004.[105]

Hybrid electric drive technologies provide power to the driving wheels by operating off the premise of using an internal combustion engine to power highly efficient electric motors, rather than less efficient hydraulic transmission. The use of this technology yields significant increases in efficiency. Therefore, incorporating hybrid electric technology into a mature fleet such as the HEMTT creates opportunities in fuel savings and simplified crew maintenance while maintaining the systems original performance capabilities. However, the fuel benefits in hybrid electric powertrains occur during stop-and-go situations rather than extended, convoy type driving conditions. This fact could present problems in Army applications. The Army sees this new technology platform as an opportunity to decrease the fleet's consumption, which decreases the logistical burden of transporting fuel as well as decrease the Army's dependence on fuel. At the same time, this increases the safety of logistic soldiers delivering fuel as well as combat soldiers providing convoy force protection.

An additional opportunity this technology provides over standard internal combustion engine powered HEMTTs is exportable power. This benefit allows the system to conduct silent watch and provide up to 100kW of power without the additional restriction of a trailer-mounted generator. In certain situations where the generator provides temporary power, this platform can replace the requirement with clean, quiet, and efficient on-board power generation. This is a key factor in military systems where the power requirements frequently exceed the power production capability of an engine-mounted generator. This in turn, removes an additional fuel-consuming burden from the logistics support structure. However, the drawback of this new technology is the drastic changes in how the powertrain functions and how the Army manages this platform. There

---

[105] Technology demonstrators are used for further testing and evaluation of a platform either by defense contractors or by DOD research and development organizations. The objective is to assess the utility of near-term deployable solutions, which respond to an endorsed military need.

are changes in repair procedures as well as powertrain components; transmissions, driveshaft, and transfer cases. Without nesting the implementation of this technology throughout the life cycle management process, the dollars from fuel savings are quickly lost.

A study conducted by Oshkosh and the Army on the A3 revealed a twenty percent increase in fuel efficiency over the newest A4 version.[106] Depending on the type of driving conditions, this equates to approximately one mile per gallon efficiency increase. This might seem like a trivial amount; however, a one mpg improvement over 5,000 miles can save more than 5,100 gallons of fuel in one year. This becomes significant if the Army utilizes this technology in the entire fleet of 17,000 HEMTTs. If each of these vehicles faced the same driving conditions and mileage while deployed, this equates to a savings of 85 million gallons. Small increases in efficiency across an entire fleet could have substantial benefits not only to the Army and its logistics but also in the DOD's dependence on fuel. Additionally, this amount does not take into account the more than 9,000 A0 and A2 versions, which have less efficient, legacy engines. Both of these earlier versions are scheduled to complete the extended service program bringing them up to the same standards as a newly built A4 including the most efficient engine available. This illustrates an even greater potential than the previously calculated 85 million gallon fuel savings as well as an opportunity to upgrade these to A3 rather than A4 versions.

The long-term development of the hybrid electric A3 intends to meet all the same requirements as each of the versions in the HEMTT family but with increased fuel efficiency and

---

[106] Nader Nasr, "Electric Drive Approach to Mobile Power Platforms" (presentation at the 2007 Joint Service Power Expo, San Diego, CA, April 24, 2007), http://www.dtic.mil/ndia/2007power/ /2007power.html (accessed December 20, 2010).

- The sequence of the versions of the HEMTT does not correspond to their production date. There are four versions still in the Army inventory: A0, A2, A3, and A4. The older versions, A0 and A2, have legacy engines, which are less fuel-efficient, and neither includes updated computer control and monitoring systems as the A3 and A4 do. The A3 and A4 both contain the most fuel-efficient engines but Oshkosh Defense built them for different requirements. The A3 diesel electric hybrid became a technology demonstrator prior to the fleet upgrades and the production of the A4. Source: Oshkosh Defense website: http://www.oshkoshdefense.com/

exportable power. In the short term, Oshkosh aims at replacing the M1120 model, which includes an enhanced load handling system capable of lifting and transporting standard Army flat-racks. There is no data available to illustrate whether the Army intends to use the A3 as a complete fleet replacement or procure it for specific niche requirements but this study assumes an entire fleet replacement in order to arrive at the maximum potential fuel savings.

This case study makes three assumptions in order to calculate the potential fuel savings of this technology. First, this case study assumes only the brigade combat teams who deployed in FY2007 would have their HEMTTs replaced with the hybrid electric A3s. Second, is that each of the brigade combat teams' HEMTTs traveled 5,000 miles per month in an expeditionary environment. In a mature theater, this is a valid assumption when considering the multiple forward operating bases relocations during a twelve to fifteen month deployment. Additionally, this value creates an average between the systems that never leave the forward operating bases and those that conduct frequent convoys throughout Iraq. The final assumption is that brigade combat teams brought their organic vehicles or assumed control of the outgoing units systems that they replaced in theater. This assumption facilitates the approximation of the number of HEMTTs in theater.

In order to support this monograph's hypotheses, these next paragraphs address each of the four focused questions. The first question is how many hybrid electric HEMTTs are required to support the predetermined troop strength and the number of forward operating bases? According to the DA G4, there is no historical data on the number of HEMTTs in theater in FY2007 to use to calculate this value. Nevertheless, an acceptable alternative to find this value is to analyze the units that deployed in 2007. The seventeen brigade combat teams' Modification Table of Organization and Equipment (MTOE) enables an estimate of 1,362 HEMTTs in Iraq in

2007.[107] However, this value delimits the platforms in the non-brigade combat team units, such as those assigned to the Sustainment Command (Expeditionary). Using only those HEMTTs derived from the brigade combat teams MTOEs enables a pattern analysis of common usage and driving conditions found in these units. Capturing data from any additional units outside the brigade exceeds the scope of this study.

The second focused question is what is the overall fuel decrease following the incorporation of the hybrid electric HEMTTs? Based on data collected from the Hybrid-Electric Vehicle Experimentation and Assessment (HEVEA) of the A3 and A4 versions of the HEMTTs, Oshkosh and the Army computed a 5,187 gallon per year fuel savings.[108] This value represents the fuel savings for one A3 version used in place of an A4. Replacing the 1,362 HEMTTs deployed in 2007 with the hybrid electric versions leads to a savings of 7.06 million gallons per year.

The third focused question is what is the overall decrease in fuel cost? The data calculated from this question presents itself in two ways: dollars per gallon from the DLA-Energy price guide and dollars per gallon using the Fully Burdened Cost of Fuel (FBCF) analysis. The first, DLA-Energy's cost of $2.21 per gallon and the second is the FBCF of $14.13 per gallon.[109] Both of these values represent the cost of fuel during FY2007, which follows the intent of this study. Calculating both values leads to the DLA fuel cost savings of $15.6 million and the FBCF cost savings of $99.8 million. Since the FY2007 budget for operations in OIF was $131.2 billion, these cost savings account for a 0.01 and a 0.08 percent reduction in the DOD's FY2007 budget.

---

[107] Calculated from an MTOE analysis on the Force Management System (FMS) website for IBCT, HBCT, CAV, and SBCTs. https://fmsweb.army.mil/unprotected/splash/ (accessed on January 30, 2011).

[108] Kozera, Report to Congress, Hybrid Electric Powertrains for Tactical Wheeled Vehicles.

[109] Army Environmental Policy Institute, *Energy and Water Costing Methodology,* vii.

The fourth focused question is what is the reduction in the number of 5000-gallon tankers no longer needed to deliver fuel? The data calculated from this question produces a value for the reduction for an entire year. This study uses the Army's standard fuel transport truck to illustrate the impact on the use of Army soldiers, not locally contracted transport. Using the amount of fuel saved from incorporating the hybrid electric HEMTTs in place of those in the brigade combat team, leads to the possibility of 1,412 tankers per year removed from roads of Iraq. According to a casualty factor study conducted by the Army Environmental Policy Institute in 2009, this reduction represents the probability that two soldiers may not have died delivering fuel in 2007.[110] This value sufficiently illustrates the reduction to soldiers, which partially supports this monograph's hypotheses.

The HEMTT A3 is a technological fuel savings breakthrough that offers the Army one of many technology options to decrease fuel use. This technology aims at a high pay off target; the cost of delivering fuel to the systems that consume fuel. The application of this technology presents a threefold benefit: increased operational effectiveness, reduced logistics, and reduced budgetary costs. What the implementation of this technology does first at the operational level is decrease operational risk to those consuming the fuel as well as those delivering and protecting it. Secondly, the reduced logistics stem from decreasing the need to store fuel used to power HEMTTs as well as the large amount of fuel used to deliver fuel. Finally, when using the FBCF analysis, which even in a mature theater appears to be a low approximation, an enterprise approach to this technology quickly decreases the DOD budget for fuel.

Even though hybrid electric drives have potential in weaning the Army off its heavy thirst for fuel, three reasons arise that stand in the way of its implementation. First, the cost of implementing this technology is significant. In order to outfit the existing fleet of HEMTTs, all 17,000 platforms would have to be replaced by new models or upgrading to the A3 version. The

---

[110] Army Environmental Policy Institute, *Sustain the Mission Project: Casualty Factors*, 6.

culture in the Army and the senior leadership in the DOD do not view this as a technology imperative. The second reason is this technology is not the golden egg; the big payoff for implementing a technology and seeing a drastic decrease in fuel consumption. The HEMTT A3 only gives a modest fuel savings of one mile per gallon, which when looked at on a individual platform basis, makes it tough to justify the required funding. Finally, the Army bases the procurement of military platforms, especially logistic systems, off capability factors. This means the Army weighs the importance of carrying a payload higher than fuel efficiency.

The reasons listed above are convincing arguments on why not to implement hybrid electric powertrains. The use patterns of these systems drive the efficiency ratings, which leads to some specific applications where this technology could benefit over other alternatives. For instance, the opportunity does exist to implement the A3 version in combat aviation brigades where the fleets are smaller and the vehicle use closely resembles stop and go driving. In addition, the Army should allow additional testing for niche use of the hybrid electric drive platforms where hotel power, or power requirements at idle, offset the minimal fuel savings.

Implications

The calculations in these last case studies do not present a compelling argument that the cost of fuel and the continued high consumption rates present a monetary and oil national security risk. Furthermore, the data shows that only an extremely small portion of the DOD's budget decreases, less than one percent, if the Army were to implement these technologies. Appendix 2 presents the figures from the case study calculations that support these assertions. As such, cost would not be why the DOD would want the Army to implement any new emerging technologies. However, what the data does support is the hypotheses in this monograph as well as present evidence that demand the implementation of new technologies. Reflecting on the summarization of the data from the case studies presents three implications, which support the further use of emerging fuel technologies.

The first significant implication that arises from the summarization of these case studies

is that the number of tankers delivering fuel across Iraq in 2007 would have dropped by 2,993.

This is nearly a 1:1 ratio of implemented emerging technologies to tankers removed from the

roads in one year: a powerful reason for the Army to continue to incorporate fuel savings

technologies. In addition, if each fuel convoy consisted of 16 fuel tankers, then 187 convoys per

year would not have faced the dangers of traveling across Iraq. This also means that with a two-

soldier crew per fuel tanker, the Army would have had 5,986 additional soldiers taken off

deployment orders or redirected to other theater missions. These crews represent the end strength

of an entire brigade combat team. There are layers of second and third order affects that occur due

to this troop reduction. As such, these affects could have significantly increased the combat

"tooth" and decreased the logistical "tail" during Operation Iraqi Freedom.

There is a risk in understating the impact that the decreased number of soldiers in Iraq has

on the DOD's deployment requirement. First, without the 5,986 soldiers in theater, the DOD

could realize decreased personnel cost for force protection, transportation, and maintenance

personnel associated with fuel delivery. Second, the DOD could realize decreased fuel usage for

force protection and transportation equipment, a second order affect of implementing these

technologies. Third, the DOD could realize additional theater wide fuel consumption decreases

due to the reduction in the sustainment requirement for these logistics soldiers. If the Army chose

not to replace these soldiers with combat personnel, there would be fewer mouths to feed, less

water to heat and purify, and less air conditioning and heating to supply.

The second implication that arises from this monograph is that the number of soldiers

killed would have decreased by four. The data proves that it is beneficial to look at increasing the

use of these technologies in order to decrease the tankers required to convoy fuel each year. This

leads to a lower probability of soldiers killed. The evidence of this lower probability is in the ratio

of the number of convoys to the probability of a soldiers being killed delivering fuel. This ratio is

24: 1. Statistically, there is one soldier killed for every 24-fuel convoys. Therefore, a decrease in

2,993 tankers stemming from the implementation of these emerging fuel technologies directly correlates to four deaths avoided in FY2007.[111] This presents an opportunity to decrease additional deaths from the 5,133 fuel convoys in FY2007, which delivered over 500 million gallons of fuel. Furthermore, the AEPI study asserts that the resupply of fuel and water accounted for 2,858 soldiers and contractors either killed or wounded in Iraq from 2003 to 2008, which substantiates the aggressive implementation of emerging fuel technologies.[112]

If this monograph seems highly selective in the choice of which emerging technologies to analyze, then it mirrors the Army and DOD's approach to identifying, prioritizing, sharing, implementing, and institutionalizing fuel reduction technologies. As such, the third implication derived from the case studies is the services' lack of efficient emerging fuel technologies management and procurement processes. Until July 2010, no office existed to merge the disparate attempts by the service to implement fuel reduction technologies and even if the technology was recognized, it was bogged down in a slow, inefficient procurement process. The slow, albeit sometimes necessary, procurement processes seems to frustrate the Army's senior leaders evident in General Peter W. Chiarelli's comments in a recent interview. When asked how long it would have taken the Army to develop and implement a foam spray recently used to cover and insulate tents in Iraq and Afghanistan, he stated, "if we had developed it, it would have taken twenty years to field it."[113] The DOD has yet to create a solution to the long, arduous procurement process because historically, it yields a solution vetted and nested through each of the Army's organizational structures. However, in a combat environment where implementing fuel savings

---

[111] Army Environmental Policy Institute, *Sustain the Mission Project: Casualty Factors*, i.

[112] Ibid., 3.

[113] Lisa Daniel, "Service Leaders Discuss Way Forward on Energy." *American Forces Press Service*, October 13, 2010, http://www.defense.gov/news/newsarticle.aspx?id=61265 (accessed March 9, 2011).

technologies save lives, the need arises for an enterprise approach to fielding these technologies in a more expedited manner.

Summary

The above implications recommend further study and oversight into the use of emerging fuel technologies early in the procurement process. In addition, these implications hint at a culture change required in senior leadership in order to create the environment where fuel savings is a top priority. To date, each of the services, their program executive offices, and the DOD's new office of operational energy plans and programs continue with their respective stove-piped approach to incorporating energy solutions.[114] Since budgets are what drive change, how would these offices' approach change if the FBCF were closer to $100 per gallon? In the case studies presented here, this equates to a $1.5 billion cost for fuel out of a $131 billion defense budget in FY2007. Crossing the one billion dollar mark could be a public perceptions trigger, which could accelerate change throughout the DOD.

Conclusion

The original intent of this monograph was to illustrate how the use of three emerging fuel technologies could reduce what the Army spends annually on fuel. What the study found instead was that, implemented piecemeal, these technologies could not reduce the DOD's hundred billion dollar budget by more than a half of a percent. Instead, the implications of using these technologies illustrated a dire need for the Army to take an enterprise approach toward emerging fuel technologies. As stated in this monograph's thesis, the benefit of combining multiple

---

[114] Lisa Daniel, "New Office Aims at Reducing Military's Fuel Usage," *American Forces Press Service*, July 10, 2010, http://www.defense.gov/News/NewsArticle.aspx?ID=60131 (accessed March 9, 2011).

technologies across the Army leads to reduced fuel consumption and reduces the deployed logistic requirements.

In order to support the thesis, the monograph focused on the hypothetical implementation of three technologies in place of existing platforms known for their heavy fuel consumption. The analysis began with calculating the number of these new technologies required to replace the existing technologies. This confirmed that a large logistical burden did not accompany this new technology. Second, the analysis determined the fuel savings from using the new technology. This revealed whether the fuel savings validated procurement of the new technology. Third, the analysis contained in the case studies facilitated the calculation of the yearly decrease in the cost of fuel for the Army. This quantified the technology in order to support part of each of this monograph's hypotheses. Finally, the analysis closed with calculations on the number of fuel tankers removed from the road following the use of the new technology. This intended to be evidence for how small investments in emerging fuel technologies significantly reduce the risk to soldiers. This data also fed into the calculation of the number of deaths avoided if the Army implemented these technologies.

Following the analysis the monograph finds that in each of the three cases the DOD budgetary reductions only account for less than one half of one percent. What the analysis did bring to light was the most profound data point from the analysis of the emerging fuel technologies. Combined, the three case studies illustrated the potential to reduce the number of 5000-gallon tankers required to deliver fuel by 2,993. This correlates to reduction in the number of soldiers driving and protecting these tankers on the roads of Iraq by more than 6,000. Once this benefit of using these technologies came to light, the monograph found that even on such a small scale as these case studies, four soldiers could avoid death while supporting fuel convoys.

It follows that the data for tanker reduction and a reduced risk to soldiers both illustrate the need for the DOD to take an enterprise approach toward the implementation of fuel technologies. In order to do this, researchers must conduct studies illustrating how emerging fuel technologies reduce the cost of fuel related force structure. As shown in a Defense Science Board report, reducing the Army's fuel demand in one year reduces the underlying personnel and infrastructure cost of fuel delivery.[115] The examples the study used was the transparent costs associated with buying $200M worth of fuel in 2001. When including the personnel force structure cost, delimiting equipment and training, the cost of that $200 million worth of fuel balloons to $3.2 billion. This means that it is 16 times more expense to deliver fuel, as it is to purchase it. Budget planners and technology research and developers do not include this when calculating the startup cost and recuperation period.[116]

The calculations of fuel savings, cost savings, and deaths avoided are additive with the use of each new emerging fuel technology. There is no single solution to cutting the Army's fuel umbilical cord, but the improvements, which arise from logistical infrastructure reductions, allow a budgetary shift of resources to other critical areas requiring attention. In order to change the Army's perception of energy technologies, the DOD must implement an enterprise approach to embracing emerging fuel technologies. Doing this decreases the cost of fuel on a national level while reducing risk to the soldiers who deliver and protect this vital resource.

---

[115] U.S. Office of the Under Secretary of Defense, *More Capable Warfighting through Reduced Fuel Burden*, 39.

[116] Ibid.

Appendix 1: Today's Top 10 Battlefield Fuel Users

SWA scenario using current Equipment Usage Profile data

Of the top 10 Army battlefield fuel users, only #5 and #10 are combat platforms

1. TRUCK TRACTOR : LINE HAUL C/S 50000 GVWR 6X4 M915

2. HELICOPTER UTILITY: UH-60L

3. TRUCK TRACTOR: MTV W/E

4. TRUCK TRACTOR: HEAVY EQUIPMENT TRANSPORT (HET)

5. TANK COMBAT FULL TRACKED: 120MM GUN M1A2

6. HELICOPTER CARGO TRANSPORT: CH-47D

7. DECONTAMINATING APPARATUS: PWR DRVN LT WT

8. TRUCK UTILITY: CARGO/TROOP CARRIER 1-1/4 TON 4X4 W/E (HMMWV)

9. WATER HEATER: MOUNTED RATION

10. HELICOPTER: ATTACK AH-64D

Source: Recreated by author from U.S. Office of the Under Secretary of Defense for Acquisition, Technology, and Logistics. *Defense Science Board Report on More Capable Warfighting through Reduced Fuel Burden* (Washington D.C. 2001), 43, http://www.acq.osd.mil/dsb/reports/ ADA392666.pdf (accessed August 24, 2010).

Appendix 2: Case Study Results

Case Study Findings Per Year	Microgrid[117]	Solar/Wind[118]	HEMTT EV[119]	Totals
Number of Systems	1,505[120]	567[121]	1,362[122]	3,434
Focused Questions				
Fuel Savings (Millions of gallons)	3.3[123]	4.6[124]	7.1[125]	12.0
DLA Cost Savings (Millions)[126]	$ 7.3	$ 10.2	$ 15.6	$ 33.1
FBCF Cost Savings (Millions)[127]	$ 46.6	$ 65.1	$ 99.8	$211.5
# of 5000-Gallon Tankers Decreased	660	921	1,412	2,993
Total Convoys Removed from the Roads	41	58	88	187
Lives Saved from Decrease in Convoys[128]	1	1	2	4
Cost of Technology (Millions)	$ 150	$ 85	$ 272	$ 507

---

[117] Teri Hall, Lockheed Martin, "Microgrid Development for Tactical Operations" (presentation at the 2009 Joint Service Power Expo, New Orleans, LA May 5, 2009).

[118] See the following websites for brief descriptions on possible solar and wind emerging technologies: http://www.windtamerturbines.com/windtamer-turbines/; http://www.skybuilt.com /products/products_skystation.htm (accessed February 15, 2011).

[119] For further information about Oshkosh's HEMTT A3 Diesel-Electric see the following website: http://www.oshkoshdefense.com/products/12/hemtt-a3-diesel-electric (accessed February 15, 2011).

[120] Gregory Bowie, e-mail message to author, February 10, 2011.

[121] Based on the number of five kW and ten kW generators in a brigade combat team calculated from an MTOE analysis on the Force Management System (FMS) website for IBCT, HBCT, CAV and SBCTs. https://fmsweb.army.mil/unprotected/splash/ (accessed on January 30, 2011).

[122] Calculated from an MTOE analysis on the Force Management System (FMS) website for IBCT, HBCT, CAV, and SBCTs. https://fmsweb.army.mil/unprotected/splash/ (accessed on January 30, 2011).

[123] In order to calculate a consumption factor of 0.83 gallons per hour for the five kW (0.57 gallons per hour) and ten kW (1.09 gallons per hour) generators in Iraq, the author took the average of two generators combined assuming an even distribution of each in theater.

[124] U.S. Department of Defense, Standard Family of Mobile Electric Power Generation Sources General Description Information and Characteristics Data Sheets Handbook, MIL-HDBK-633A, January 26, 2010, http://www.everyspec.com (accessed on January 30, 2011).

[125] Kozera, Report to Congress, *Hybrid Electric Powertrains for Tactical Wheeled Vehicles*.

[126] Army Environmental Policy Institute, *Energy and Water Costing Methodology*, vii.

[127] Ibid.

[128] Army Environmental Policy Institute, *Sustain the Mission Project: Casualty Factors*, 6.

Bibliography

Books

Brown, Charles E. *World Energy Resources*. Berlin, Germany: Springer, 2002.

Brune, Michael. *Coming Clean: Breaking America's Addiction to Oil and Coal*. San Francisco, CA: Sierra Book Club, 2008

Chester, Edward W. *United States Oil Policy and Diplomacy, A Twentieth-Century Overview*. Wesport: Greenwood Press, 1983.

Corsi, Jerome R., and Craig R. Smith. *Black Gold Stranglehold: The Myth of Scarcity and the Politics of Oil*. Nashville, TN: WND Books, 2005.

Deese, David A., and Joseph S. Nye. *Energy and Security*. Cambridge: Ballinger, 1981.

Klare, Michael T. *Resource Wars: The New Landscape of Global Conflict*. New York, NY: Henry Holt and Company, 2001.

Sandalow, David. *Freedom from Oil, How the Next President Can End the United States' Oil Addiction*. New York: McGraw-Hill, 2008.

Stoff, Michael B. *Oil, War, and American Security*. New Haven: Yale University Press, 1980.

Vansant, Carl. *Strategic Energy Supply and National Security*. New York: Praeger, 1971.

Waddell, Steve R. *United States Army Logistics, The Normandy Campaign, 1944*. Westport: Greenwood Press, 1944.

Weintraub, Sidney, ed. *Energy Cooperation in the Western Hemisphere*. Center for Strategic and International Studies. Washington, DC: CSIS Press, 2007.

Articles

Brown, Lynda. "Bulk Services Reach Farther with Afghanistan Transportation Contract." *Fuel Line*, Defense Energy Supply Center, January 2009.

Daniel, Lisa. "New Office Aims at Reducing Military's Fuel Usage." *American Forces Press Service*, July 10, 2010. http://www.defense.gov/News/NewsArticle.aspx?ID=60131 (accessed March 9, 2011).

Daniel, Lisa. "Service Leaders Discuss Way Forward on Energy." *American Forces Press Service*, October 13, 2010. http://www.defense.gov/news/newsarticle.aspx?id=61265 (accessed March 9, 2011).

Daniel, Lisa. "Department Hailed as Leader in 'Green' Movement." *American Forces Press Service*, April 20, 2010. http://www.defense.gov/news/newsarticle.aspx?id=58831 (accessed August 5, 2010).

Fine-Maron, Dina. "Can a Microgrid Protect U.S. Troops in Afghanistan?" *Scientific American*, December 22, 2010. http://www.scientificamerican.com/article.cfm?id=can-a-microgrid-protect (accessed February 10, 2011).

Foster, Taft and Thomas Klier. "Raising automotive fuel efficiency." *Chicago Fed Letter* no. 266, The Federal Reserve Bank of Chicago. September 2009. http://www.chicagofed.org/webpages/publications/chicago_fed_letter/2009/september_266.cfm (accessed December 10, 2010).

Harrington, Caitlin. "U.S. Army launches initiative to reduce fuel consumption." *Jane's Defense Weekly*. October 15, 2008. http://search.janes.com/Search/printFriendlyView.do?docId=/content1/janesdata/mags/jdw/history/jdw2008/jdw38003.htm@current (accessed August 5, 2010).

Khalil, Ghassan, Major Christine E. Allen, and Michael Pozolo. "Hybrid-Electric Vehicle Experimentation and Assessment (HEVEA) Program Supports the Army's Need for Increased Power Demands." *Army AL&T Magazine,* January-March 2009.

Kyzer, Lindy. "Army turning trash into energy in Iraq." *Army News Service*. June 19, 2008. http://www.army.mil/-news/2008/06/19/10194-army-turning-trash-into-energy-in-iraq (accessed October 25, 2010).

Magnuson, Stew. "Ensuring No One Pays the 'Ultimate Price' for Fuel Becomes New Goal." *National Defense*, August 2009.

Tiron, Roxana. "Fuel's Full Burden." *Sea Power* 53, no. 3 (March 2010): 30-32. *Academic Search Complete*, EBSCO*host* (accessed March 28, 2011).

U.S. Department of Defense. Office of the Assistant Secretary of Defense (Public Affairs) "DoD Announces Units for Next Operation Iraqi Freedom Rotation." June 27, 2006. http://www.defense. gov/releases/release.aspx?releaseid=9770 (accessed December 30, 2010).

U.S. Department of Defense. Office of the Assistant Secretary of Defense (Public Affairs). News Transcript. "Remarks by Secretary Gates at the Command and General Staff College." http://www.defense.gov/transcripts/transcripts.aspx?transcriptsID=4623 (accessed August 5, 2010).

Presentations

Hall, Teri. "Microgrid Development for Tactical Operations." Lockheed Martin Corporation presentation, 2009 Joint Service Expo, New Orleans, LA, May 5, 2009. http://www.dtic.mil/ndia/2009power/ (accessed December 20, 2010).

Nasr, Nader. "Electric Drive Approach to Mobile Power Platforms." 2007 Joint Service Power Expo, San Diego, CA, April 24, 2007. http://www.dtic.mil/ndia/2007power/2007 power.html (accessed December 20, 2010).

Richard, Paul. "Mobile Electric Power for Today and Tomorrow." Department of Defense Project Manager Mobile Electric Power, 2009 Joint Service Expo, New Orleans, LA, May 5, 2009. http://www.dtic.mil/ndia/2009power/ (accessed December 20, 2010).

Public Documents

Global Security. "Iraq Facilities." http://www.globalsecurity.org/military/facility/iraq.htm (accessed December 21, 2010).

Kramer, Thomas D. "Addicted to Oil: Strategic Implications of American Oil Policy." *Carlisle Papers in Security Strategy*. U.S. Army War College. Carlisle, PA: USA, May 2006.

Rivera, Allen. "Cost Benefit Analysis of integrated COTS Energy-Related Technologies for Army's Force Provider Module." Master's Thesis, Naval Postgraduate School, September 2009. http://handle.dtic.mil/100.2/ADA509914 (accessed November 12, 2010).

U.S. Congress. House Committee on Armed Services Readiness Subcommittee. *Testimony of Chris Dipetto*. Office of the Deputy Under Secretary of Defense (Acquisition & Technology). March 13, 2008.

U.S. Congress. Senate. *Duncan Hunter National Defense Authorization Act for Fiscal Year 2009*. S.3001. 110th Cong., *Congressional Record Summary*, The Library of Congress. (October 14, 2008) http://thomas.loc.gov (accessed May 17, 2011).

U.S. Department of Defense. *Defense Energy Support Center 2009 Fact Book*. http://www.desc. dla.mil/DCM/Files/FY09%20Fact%20Book%20(8-10-10).pdf (accessed October 20, 2010).

U.S. Department of Defense. *Fiscal Year 2009 Budget Request Summary Justification February 4, 2008*. http://comptroller.defense.gov/defbudget/fy2009/Summary_Docs/FY2009_ Budget_ Slides.pdf (accessed March 24, 2011).

U.S. Department of Defense. *Quadrennial Defense Review 2010*. Washington, D.C: Government Printing Office, 2010. http://www.defense.gov/qdr/images/QDR_as_of_12Feb10_1000 .pdf (accessed May 17, 2011).

The White House. *National Security Strategy 2010*. Washington, DC: Government Printing Office, May, 2010. http://www.whitehouse.gov/sites/default/files/rss_viewer/national_ security_strategy.pdf (accessed September 24, 2010).

Reports

Army Environmental Policy Institute. *Sustain the Mission Project: Casualty Factors for Fuel and Water Resupply Convoys, Final Technical Report, September 2009*. by David S. Eady, Steven B. Siegel, R. Steven Bell, and Scott Dicke. http://www.aepi.army.mil/ docs/whatsnew/SMP_Casualty_Cost_Factors_Final1-09.pdf (accessed December 20, 2010).

Army Environmental Policy Institute. *Sustain the Mission Project: Energy and Water Costing Methodology and Decision Support Tool: July 2008*. by Steve Siegel, Steve Bell, Scott Dicke, and Peter Arbuckle. http://www.aepi.army.mil/docs/whatsnew/SMP2_ Final_Technical_Report.pdf (accessed December 20, 2010).

Brookings Institute. *Iraq Index: Tracking Variables of Reconstruction & Security in Post-Saddam Iraq.* by Michael E. O'Hanlon and Ian Livingston. Washington, DC: Brookings, December 9, 2010. http://www.brookings.edu/iraqindex (accessed 20 December 2010).

Congressional Research Service. *Department of Defense Fuel Costs in Iraq.* by Anthony Andrews and Moshe Schwartz, RS22923. Washington, DC: U.S. Library of Congress, 2008. http://www.policyarchive.org/handle/10207/bitstreams/18707.pdf (accessed December 20, 2010).

Congressional Research Service. *Department of Defense Fuel Spending, Supply, Acquisition, and Policy.* by Anthony Andrews, R40459. Washington, DC: U.S. Library of Congress, 2009. http://assets.opencrs.com/rpts/R40459_20090922.pdf (accessed October 15, 2010).

Crane, Keith., Andreas Goldthau, Michael Toman, Thomas Light, Stuart E. Johnson, Alireza Nader, Angel Rabasa, and Harun Dogo. *Imported Oil and U.S. National Security.* Santa Monica, California: RAND Corporation, 2009. http://www.rand.org/pubs/technical_reports/ (accessed September 24, 2010).

Crowley, Thomas, Tanay Corrie, David Diamond, Stuart Funk, Wilhelm Hansen, Andrea Stenhoff, and Daniel Swift. *Transforming the Way DOD Looks at Energy: An Approach to Establishing an Energy Strategy.* FT602T1. LMI Government Consulting, April 2007. http://www.dtic.mil/cgi-bin /GetTRDoc?AD=ADA467003&Location=U2&doc=Get TRDoc.pdf (accessed November 23, 2010).

Deloitte. *Energy Security, America's Best Defense,* November 2010. http://www.deloitte.com/view/en_US/us/Industries/oil- gas/ (accessed May 24, 2011).

Hayward, Steven F., Mark Muro, Ted Nordhaus, Michael Shellenberger. *Post-Partisan Power.* http://www.aei.org/paper/100149 (accessed October 25, 2010).

JASON. *Reducing DOD Fossil-Fuel Dependence, September 2006.* JSR-06-135. Paul Dimotakis, Robert Grober, Nate Lewis. Prepared by JASON in cooperation with the Office of the Deputy Under Secretary of Defense (S&T). http://www.fas.org/irp/agency/dod/jason/fossil.pdf (August 24, 2010).

U.S. Department of the Army. *Army Truck Program: Tactical Wheeled Vehicle Acquisition Strategy Report to the Congress.* Washington, DC: June 2010. www.defenseindustry daily.com/2010_US_Army_Tactical_Wheeled_Vehicle_Acquisition_Strategy.pdf (accessed February 15, 2011).

U.S. Department of Defense. *Annual Energy Management Report: Fiscal Year 2006.* January 2007. http://www.acq.osd.mil/ie/energy/energymgmt_report/fy06/FY_2006_Annual_Report_Narrative.pdf (accessed January 10, 2011).

U.S. Department of Defense. Office of the Under Secretary of Defense for Acquisition, Technology, and Logistics. *Defense Science Board Report on More Capable Warfighting through Reduced Fuel Burden.* Washington, DC: 2001. http://www.acq.osd.mil/dsb/reports/ADA392666.pdf (accessed August 24, 2010).

U.S. Department of Defense. Office of the Under Secretary of Defense for Acquisition, Technology, and Logistics. *Defense Science Board Task Force Report on DOD Energy*

Strategy: "More fight - Less fuel." Washington, DC: 2008. http://www.acq.osd.mil/dsb/reports/ADA477619.pdf (accessed August 24, 2010).

U.S. Government Accountability Office. *Defense Management: DOD Needs to Increase Attention on Fuel Demand Management at Forward-Deployed Locations.* GAO-09-300. February 2009. http://www.gao.gov (accessed August 24, 2010).

U.S. Government Accountability Office. *Defense Management: Overarching Organizational Framework Could Improve DOD's Management of Energy Reduction Efforts for Military Operations, 2008.* GAO-08-523T. http://www.gao.gov/htext/d08523t.html (accessed August 24, 2010).

Technical Manuals

Headquarters, Department of the Army and Air Force. *TM-9-2330-394-13-P M1082 Series 2 ½ Ton Light Tactical Vehicle Trailer (LMTVT) and M1095 Series 5 Ton Medium Tactical Vehicle Trailer (MTVT) Manual.* Washington, DC: Government Printing Office, January 2005.

U.S. Department of Defense. *Standard Family of Mobile Electric Power Generation Sources General Description Information and Characteristics Data Sheets Handbook*, MIL-HDBK-633A, January 26, 2010. http://www.everyspec.com (accessed on January 30, 2011).

www.ingramcontent.com/pod-product-compliance
Lightning Source LLC
Chambersburg PA
CBHW081224170526

45165CB00009B/2941